WHEEL
of FORTUNE

Work and Life in the Age of Falling Expectations

Jamie Swift

Between the Lines

Published by:
Between The Lines
720 Bathurst Street, #404
Toronto, Ontario M5S 2R4
Canada

Design: Gordon Robertson
Cover illustration: David Laurence
Printed in Canada

Between The Lines gratefully acknowledges financial
assistance from the Canada Council, the Canadian Heritage
Ministry, the Ontario Arts Council, and the Ontario
Ministry of Culture, Tourism, and Recreation.

Canadian Cataloguing in Publication Data

Swift, Jamie, 1951-
Wheel of fortune : work and life in the
age of falling expectations

Includes index.
ISBN 0-921284-88-8 (bound)   ISBN 0-921284-89-6 (pbk.)

1. Work.   2. Labor market – Ontario – Windsor.
3. Labor market – Ontario – Kingston.   I. Title

HD8106.5.05S8 1995   331.1'0971   C95-930761-3

CONTENTS

Preface   VII

1 Introduction:
   The Age of Falling Expectations   I

2 Windsor:
   "You Could Quit at Ford's in the Morning . . ."   19

3 Kingston:
   ". . . With Not Much in Between"   41

4 The Training Gospel and the Army of Servants   70

5 Public Purpose, Know-how, and the War of Production   94

6 Private Troubles in the Border City   115

7 Collars of Many Colours in the Limestone City   147

8 Flexiworkers and the Future:
   "All That Is Solid Melts into Air"   174

9 Work, Time, and the Wheel of Fortune   209

   Notes   249
   Index   269

PREFACE

I FIRST THOUGHT of writing this book after reading a report called *Good Jobs, Bad Jobs: Employment in the Service Economy*, published by the Economic Council of Canada just before the recession that began in 1990. I became interested in exploring the sometimes dramatic, sometimes subtle changes that are occurring in the world of work and the labour market, and in looking at the impact of those changes on people's everyday lives.

In a way this book is about the future as much as the present. It also contains a good deal of history. I will admit to being sceptical about the claims of writers who say a "new economy" will be restructured when workers seize the means of production by purchasing home computers. I am no futurist. I have no grand theories about Third Waves or New Industrial Revolutions or New Economies. But I do know that many Canadians are facing a precarious future as they struggle to adapt to changes just now coming into focus.

The recession of the early 1990s hit Ontario especially hard, and I wanted to find out what all the talk about jobs and joblessness, training and non-training, new economy and old meant for working-class people, families in which parents and often grandparents had held the old, apparently secure jobs. What is their present? What do they feel about their futures? What do these changes mean for their institutional expression, the trade union movement, a movement that is on the defensive?

After coming across countless references to the "new job require-ments" of the "new economy" of the 1990s I began to poke through the archives in search of clues to help make sense of the puzzle that is job training. I looked through a small fraction of the academic litera-ture on the labour market, training, skill, and social class in an attempt to understand how these elements fit together.

I also talked with people in two Ontario cities about their past work, their present jobs, and what they feel about the future. I went to Windsor because it is a manufacturing town dependent on the car industry. I also talked with people at the eastern end of Lake Ontario, where I live. If Windsor is an old-fashioned lunch-bucket town, Kingston's service economy seems to represent a step into the future. I've been guided in large part by what the people in these two cities told me.

Based as it is, at least in part, on conversations with people in two southern Ontario cities, this book has a particular focus of place. I gathered other sources of research information from the Ontario Public Archives, so some of what follows is particular to policies as they emerged in that province as well as in Ottawa. Although I may be accused of Eastern Ontario arrogance by readers from the West, or of Central Canadian presumption by readers from Atlantic Canada, I am convinced that the conditions and changes in social relations the book describes are appearing most everywhere else in urban Canada. I hope that readers in the rest of Canada will forgive any implied overgeneralizations.

This sort of effort cannot be undertaken without help. I'd like to thank librarians and archivists at the May Ball, Douglas, and Stauffer Libraries at Queen's University; people at the Archives of Ontario, the Windsor Public Library, and the Kingston Public Library. Rick Coronado, Jim Brophy, and Ron Dickson made Windsor easier to understand. Roberta Hamilton gave me the idea of looking around closer to home, and read the manuscript; so did Jerry Bickenbach, Catherine Macleod, Harvey Schachter, Ian McKay, Don Wells—and John Holmes. David Peerla offered consis-tently helpful comment and further bibliographical tidbits. I am grateful to the Canada Council which helped to support the research and writing of this book. Between The Lines, and especially BTL's Marg Anne Morrison and Pat Desjardins, made it all possible. Susan and Sonya were patient enough to tolerate dark and early mornings and (sometimes) dark moods, providing cheerful diversion. As always errors and shortcomings can be blamed on me—and of course my editor, Robert Clarke. It is his book as much as it is mine.

1

INTRODUCTION

The Age of
Falling Expectations

That feeling that you've got to be everlastingly fighting and hustling, that you'll never get anything unless you grab it from somebody else, that there's always somebody after your job, that next month or the month after they'll be reducing staff and it's you that'll get the bird.

– George Orwell, *Coming up for Air*, 1939

ENGLISH WRITER George Orwell is best known for his novels *Animal Farm* and, especially, *1984*—the book that would turn its author's name into an adjective. The word "Orwellian" conjures up the bleak vision of the future as totalitarian nightmare. "If you want a picture of the future," Orwell wrote in *1984*, "imagine a boot stamping on a human face—for ever."

Orwell was a crusader not just against totalitarianism but for what he always called "decency." He was warning against *indecent* tendencies that exist in all mass societies. But today one of Orwell's less celebrated novels seems just as pertinent—and perhaps even more so—than his better known books. *Coming up for Air*, published in 1939, was Orwell's first successful novel. He wrote it in the first person, describing the world through the eyes of a charter member of the middle class, a pudgy salesman named George Bowling who felt the need to ditch his stale job and drab life for a few days and go search out his old home town and recover his youth.

> I was down among the realities of modern life. And what are the realities of modern life? Well, the chief one is an everlasting, frantic struggle to sell things. With most people it takes the form of selling themselves—that is to say, getting a job and keeping it . . .

This passage strikes me as being just as Orwellian as anything in *Animal Farm* or *1984*—perhaps because of its familiarity. The "realities of modern life" in the mid-1990s are more than ever about a frenzied sales effort. As jobs are eroded by a potent brew of computerization, footloose factories, downsized offices, and a shrinking public sector, people compete in an increasingly global labour pool. The "realities of modern life" have gone beyond anything Orwell might have imagined, at least when it comes to the hucksterism of the market—people selling things and selling themselves.

An overstatement? Take a look through a newspaper's classified advertisement section, where notices from a growing array of training corporations have created a whole new listing called "Education" that jostles the old help-wanted columns. The help that is wanted is very often of the pizza delivery, waiter, and beauty salon variety, together with the usual come-ons for Avon representatives and other low-end commission sales jobs. Or think about it the next time you get a call around suppertime from some poor devil who's been forced into a job as a "telemarketer." These people try to talk people they'll never meet into buying stuff they don't want, or need.

Of the hundred-odd people I interviewed for this book, one comment in particular stayed with me. It came from a young man who spent his days at work staring into a computer screen in what he called a "word factory." He had previously tried his hand at telemarketing, but, he said, "I began to have more empathy for the people robbing variety stores."

Montreal may well provide a snapshot of what the future holds in a so-called "new economy." By the end of 1994, a major recession was definitely over. There was talk of a job rebound concentrated in what are called "non-traditional sectors." One company, Suburban Canadair, had added 350 jobs during the year, many of them in well-paid engineering positions. Tourism was booming, in good measure due to a new casino that had attracted nearly a million out-of-town visitors. The auto-routes were dotted everywhere with fresh new road signs: an ace-jack logo (the perfect blackjack hand) pointed potential gamblers in the right direction. An economist for the huge Desjardins credit union was impressed enough to describe the job rebound as "a little shocking."[1]

But on Christmas eve a standard newspaper article on charity hampers and free church-hall turkey dinners provided facts that were no less shocking. One in four Montreal households brings in under $10,000 annually. According to Statistics Canada, nearly half of Montreal's citizens live below the poverty line, making the city "the poverty capital of Canada."[2] Despite economic growth and record corporate profits (particularly by banks), the official unemployment rate stood stubbornly at 12.1 per cent. Grizzled men standing with their palms outstretched at the bottoms of escalators on the Metro have—along with high-profile jazz, comedy, and film festivals—become permanent fixtures on the Montreal scene. Some of the men are young, some are old. All look like veterans of life on the street.

Montreal was once a prosperous city, both manufacturing centre and the undisputed financial capital of a nation drawing its wealth from agriculture and natural resources. Indeed, in the distant past, prosperity in Canada in general came largely from the land, and from the skills and toil of the people who worked the land and the waters, who harvested or extracted the resources. But when farming and the resource industries became more and more mechanized, Canadians and newcomers moved to cities in the often desperate hope of finding work in manufacturing industries, large and small. At first manufacturing was the take-up sector that helped to absorb a flood of agricultural and immigrant labour. In the immediate postwar boom period, six out of ten Canadians worked in the goods-producing sector—on assembly-lines, on farms, in mines and forests, on the sea.

But by the middle of the twentieth century, as the automation of industrial production accelerated, people began to drift into the service sector in greater numbers. In the past forty years, nine out of every ten jobs created in Canada has been in services. More than seven out of every ten workers are employed in the service industries—all this in only one generation following the postwar boom.[3]

Today the east-coast fishery has been devastated, and the forest frontier is receding. In 1994 there were about seventy-four thousand jobs in the pulp and paper industry; by the end of the decade these jobs could decline by some fifteen to twenty thousand.[4] The manufacturing industries, already concentrated in a very small section of a very big land, have been decimated by free trade, technological change, and corporate restructuring.

Between 1966 and 1988 the goods sector that produces the newsprint, ketchup, cars, and pills grew by an average of 3.4 per cent annually. The service sector—including people who dream up ads for the cars, and who sell, insure, repair, and license them as well as clean up the emergency rooms after auto accidents—grew by 6.4 per cent every year.[5]

The Economic Council divided the service sector into three parts: non-market, traditional, and dynamic services. Non-market services are the public, welfare-state industries: nursing, teaching, social work. Traditional services are the barbering and burger-flipping and bed-making jobs that usually spring to mind when the word "service" is mentioned. Dynamic services include employment and job training agencies as well as a huge array of banking, insurance, advertising, communication, transportation, legal, and wholesale jobs. Dynamic services is where so much computer-based automation is occurring.

Personal Touch Banking units and Green Machines proliferate, spewing out cash for bank customers. The Toronto-Dominion (TD) Bank typifies how financial institutions use computer power. The TD Bank has a fistful of trade-mark-registered services with names like

4

Money Monitor, Business Window, Assetlink, Notewriter, and SPEED, which allow commercial customers to do everything from checking their loan balances to issuing commercial paper in any denomination of currency—all with personal computer terminals in their own offices. Individual customers will be able to gain access to a complete range of banking services using their telephones at home. A report by the investment bankers Salomon Brothers estimated a 50 per cent increase in the productivity of financial services when banks and other institutions deploy this kind of technology. According to Arthur Cordell and Ran Ide, financial companies "make distinct savings in the time previously spent by workers doing jobs that are now accomplished more efficiently and in less time by machines."[6]

Today most of us work in services—clerking, managing, selling, distributing, guarding, researching, consulting, caring. But technology is rapidly displacing service workers. Where will those workers end up? Is there a place for them in a new take-up sector? Will there be a new take-up sector?

It seems that in Canada the number—and type—of jobs available has always been linked to the idea of economic growth, and technological progress. By mid-1990s the economy had gone through another of its familiar cycles. In the 1980s the record unemployment of 1983 (11.8 per cent) had given way to a partial job recovery at the same time that growth soared and business boomed. But unemployment remained stubbornly high—by 1989 it had gone down to only 7.5 per cent. Twenty years earlier this would have been seen as a scandalously high level of joblessness.

Permanently high unemployment should have been no surprise; as early as 1983 the Canadian Manufacturers' Association was predicting a return to full capacity with 10 per cent fewer workers. In 1992 the CMA's chief economist warned, "Right now, there is no job in Canada that is secure."[7] Some three years later—and after what economists were declaring a full-fledged recovery—the jobless level was still hovering at around 10 per cent. The link between growth and jobs was becoming less clear.

Telecommunications, surely a vibrant sector in the "information age," are a case in point. In the United States AT&T is eliminating a third of all operators, some six thousand workers, replacing them with voice recognition technology that can respond to demands from customers. In 1993 Bell Canada said it intended to eliminate five thousand jobs from its payroll. The announcement came a few days after the company received a union bargaining position pointing out how employment in both craft and operator services had declined by 14 per cent to eighteen thousand workers in the past three years. In 1994 GTE, the largest U.S. supplier of local phone service (it also has a controlling interest in B.C. Tel and operations in Quebec), announced that it intended to cut seventeen thousand jobs within three years.[8]

Recognizing the new possibility of "jobless growth," in 1980 the heretical French Marxist Andre Gorz wrote a book on the "post-industrial revolution" entitled *Farewell to the Working Class.* Gorz argued: "The point now is to free oneself *from* work by rejecting its nature, content, necessity and modalities." Thinkers like Gorz and Ivan Illich, whose ideas are grounded in a critique of consumerism as well as in the urgency of a decaying natural environment, have realized that poverty and productivity, inequality and competitiveness, dollar stores and chic boutiques can co-exist very nicely against a background of an increasingly segmented labour market.[9]

A market in labour has long existed. Even before English philosopher Thomas Hobbes declared that "a mans [sic] Labour is a *commodity* exchangeable for benefit, or any other thing," French workers were offering themselves up for hire on Avignon's famous bridge to have their attributes inspected by curious employers. In the pre-dawn of capitalism, similar labour markets existed from Bohemia to Brittany and indeed throughout Europe. "All he had to offer was his arm or his hand, his 'labour' in other words," observed historian Fernand Braudel. "And of course his intelligence or skill."[10]

Skill: "No longer can prosperity come straight *from the ground*," the Economic Council of Canada stated just before the recession that began in 1990. "Increasingly, it must come *from the minds* of the Canadian people."[11] Canadian prosperity has always depended on both raw resources *and* the skills of the artisans, craft workers, and, yes, "ordinary workers" who have taken those resources and transformed them into at least semi-finished goods. Today these various people are labelled, with a sort of brutal honesty, "human resources." In Ottawa, what was once Health and Welfare and the Labour ministry has now become Human Resources Development. "Personnel departments" have gone into decline, replaced by "Human Resource" specialists.

This sort of vocabulary shift should give us pause to reconsider how human resources are used by the great private and public bureaucracies that deploy them. Historically we have not done so well in our treatment of natural resources: witness the state of our fisheries and the retreating forest frontier, our despoiled lakes and rivers; or the fact that the sun and rain—once regarded as blessings—have come to be seen as threats. Of course, human resources are also natural resources in the great scheme of things. Are they to be treated as disposable commodities in the Hobbesian sense? Perhaps not. In today's scheme of things, human resources are viewed as commodities that can be recycled (in other words, retrained) when they become obsolete.

With the rise of human resources has come a renewed emphasis on what those resources have to offer. Talk about the need for individual Canadians to sharpen their wits to compete in the new global economy is usually combined with an emphasis on new skills needed in the service and information-based economy. Our future prosperity, it is said, depends on how each of us performs. "Employers everywhere are demanding higher educational and skill levels," the chairman of the Business Council on National Issues told one management faculty.[12] This mantra is intoned in rote fashion. We hear it so often that in the pages that follow I shall simply refer to it as the Training Gospel. But what would happen, I wonder, if we had a perfect Canada? A place where everyone reached for the top and actually *attained* those higher skill and education levels? We can only assume that in this perfect world there would be jobs for all.

7

Robert Reich, Bill Clinton's labour secretary and a scholarly observer of labour-market trends, told us how to prepare for this capitalist future in terms that resonate strongly with the tones of class:

> People fortunate enough to have had an excellent education followed by on-the-job experience doing complex things can become steadily more valuable over time, making it difficult for others ever to catch up. In fact, their increasing advantage may extend beyond a single generation, as their extra earnings are invested in their children's education and training. Such widening divergences may be endemic to a global economy premised on high-value skills rather than on routine labour or capital.[13]

Those workers who get a good education will measure up and may get one of the good jobs at the top of the service sector and prosper, providing they are flexible and willing to retrain as the labour market shifts beneath their feet. Those who do not—or cannot— measure up will quickly be filtered down, and down—to join a growing number of unemployed and underemployed people who can only aspire to poorly paid jobs at the bottom of the service sector: society as Darwinist wringer.

During the early postwar years incomes from work or investments tended to increase equality in Canada. But since the early 1970s these "market earnings" have shown a different tendency. Labour-market *inequality* has been on the rise, producing a more polarized society: a marked division between, on the one hand, decently paying, secure (or "good") jobs and, on the other, lower paying, insecure, often part-time (or "bad") jobs.

This trend to polarization is reflected in income distribution. In 1967, 27 per cent of Canadian workers had middle-level incomes. By 1986 the figure was down to 22 per cent, with the middle being squeezed most in industrial Ontario, where there was a drop of 8 per cent. Similarly, by 1993 Statistics Canada was reporting that the 60 per cent of Canadians in the middle-income brackets had lost between 3.2 and 4.2 per cent of their incomes. Those in the top income bracket had lost only 1.2 per cent, perhaps because they benefited from tax policies skewed in their favour.[14]

The same trend was evident among families. In 1967 34.9 per cent of families reported middle-level incomes, but by 1986 the figure was 29.6 per cent—even though by then more women were contributing to the family income total. In addition, market-income poverty increased between 1973 and 1984, then dropped off slightly before rising again in the post-1989 period. Throughout this period job growth took place at an astonishingly high rate in the area of part-time, temporary, and precarious employment.

In 1990, *before* the impact of free trade and recession, the Economic Council of Canada described such changes as "fundamental and systemic." In 1993 Decima Research reported that 82 per cent of those surveyed felt it was getting more difficult to get by on hard work and merit alone.[15] "The volatility of incomes," the Economic Council of Canada concluded in its 1992 report *The New Face of Poverty*, "implies that roughly one-third of working-age Canadians face the risk of being poor at some time in their working lives."[16]

It is the adjective here that is important: " . . . *working* lives." Between 1981 and 1991 the number of households classified as *working poor* grew by 30 per cent. For unattached individuals the rise was even more startling: 57 per cent. What follows includes conversations with working people, and many of these people are close to being—or have already become—what is known as "the working poor."

In Windsor Lorraine Wilkinson sits in a smoky office at Local 195 of the Canadian Auto Workers. Housed in a huge converted discount store on Ottawa Street, 195 was the first local chartered by the United Automobile Workers when the Congress of Industrial Organizations was making its historic inroads into Ontario's auto industry in the 1930s.

Wilkinson, who had recently made her last spark plug after spending twenty-four years working at the Champion plant, can only answer my questions with a torrent of her own. "Don't you have to have a manufacturing base? A product base? Who are all these people going to service? Don't people have to actually make things to provide jobs for people to service?

"The places that traditionally hired women—Champion, Wyeth, Complax—they're all gone," Wilkinson says. GM trim could be the next to go, heading south like many of the others. "Now even college grads aren't getting jobs. It's an employer's market and they kind of dangle work in front of you. It's sad, you know, it's sad."

In Windsor in the 1990s there are still good jobs at the top end of the industrial labour market. Several thousand workers, for instance, still toil at $22 an hour producing minivans at Chrysler's highly successful plant. There are, as always, a relatively small number of good, well-paying jobs in management, consultancy, and at the high end of the service industry. But many more bad jobs have settled into the bottom of the service sector, creating a social structure that looks far different than the one that emerged during the postwar boom years, when unemployment was low.

My first inclination in writing this book was to use the image of an hourglass to describe the new economy. Pinched in the centre, the hourglass seemed to convey the idea of a shrinking middle. But an hourglass must, to fulfil its function, contain equal halves. Perhaps a pear—fat at the bottom and tapering gradually to the top—might be more appropriate, or maybe the idea of a split-level society. What is clear is that Canada is becoming more segmented and unequal, more like our neighbour to the south.*

In 1988 sociologist John Myles, who has worked (in "good" service-sector jobs) as a university professor and a consultant on contract at Statistics Canada, described the very real threat of a

---

* The comparison with the United States takes on foreboding spatial characteristics in the case of Montreal, where growth is increasingly concentrated in the suburbs, the places to which secure jobs and incomes have fled, leaving a city centre with a diminished tax base and a lot of poor people dependent on limited—and indeed reduced—public services.

"hamburger economy" of low-wage burger-flippers in the emergence of a split- or dual-labour market.[17] This translates into increased insecurity. In his book *The Culture of Contentment* the Canadian-born economist John Kenneth Galbraith argues that the United States (and with free trade and reduced state commitment to public provision the argument can be extended to Canada) has become a split-level society consisting of a financially contented elite made up of managers who staff the middle and upper levels of corporations, the rich who rely on inherited wealth, independent business operators, a large number of lawyers, doctors, engineers, scientists, accountants, and journalists, and other members of the modern professional class.

"Included also are a certain if diminishing number of what once were called proletarians," Galbraith says. He also pinpoints what he calls a fixture in modern capitalist societies like Canada and the United States—a "functional underclass." He concludes: "The normal upward movement that was for long the solvent for discontent has been arrested. The underclass has become a semipermanent rather than a generational phenomenon."[18]

In its attempt to avoid the fate of so many decaying U.S. inner cities Windsor introduced a glitzy casino—and blackjack tourism—into a forlorn downtown cluster of bingo parlours, marginal retail outlets, and abandoned shops. In 1992 the city, reeling from the effects of a shrinking industrial base and a surge in cross-border shopping, was chosen as the site for Ontario's first officially sanctioned casino. The Windsor casino opened in 1994 and, like Montreal's glitzy gambling palace, was a fast and huge success. In its first year it attracted 3.9 million visitors—3.1 million of them from outside Canada—and generated around $550 million in revenues, making it (according to the Ontario Casino Corporation) one of the world's leading casinos.

It seems that communities everywhere are searching for ways to attract visitors and their money, for tourism is the world's biggest employer. But although the visitor business may provide high-end jobs for managers, accountants, and the various consultants who for a short period of time work at turning a "concept" into a reality, the long-term jobs of cleaning pots, making beds, washing floors, and fielding reservation inquiries will always be rather lacklustre—and lacking in decent pay and benefits.

Meanwhile, at the eastern end of Lake Ontario Kingston has tied its hopes for the future to head offices and administration, software, biotechnology, and research and development—with an emphasis, as everywhere else, on education and training. Kingston's "historic" downtown is gentrified and comparatively prosperous. The Limestone City caters to a lively summer tourist business, trading on its colonial and Loyalist persona. Its economic spark plugs include a major university, a sophisticated hospital system, government offices, and prisons. The one remaining solid industrial employer is a Du Pont nylon plant. The city used "to chase smokestack industries," the director of Kingston's Economic Development Commission told me. But now its strategy "aims at people-based service industries."

In 1994 the Correctional Worker Program at St. Lawrence College of Applied Arts and Technology had 547 applications for 50 spaces. The Human Resources office of Corrections Canada is just down the street from St. Lawrence; in 1992-93 the prison managers received three thousand applications for a hundred positions.[19] Meanwhile, six thousand people were crowding three Windsor-area Canada Employment Centres in search of application forms for two hundred jobs at the Chrysler minivan plant.

Windsor and Kingston are hardly unique in this. In early 1995 just outside Toronto between fifteen thousand and twenty thousand people lined up, attracted by rumours of a possible twelve-hundred high-wage jobs that would appear if General Motors decided to add a third shift to its Oshawa assembly-line. The company gathered twenty-six thousand applications in two days, with "The Lineup" taking on enormous symbolic value in renewed discussions of the jobless recovery.

The current situation is about more than the usual surplus of labour that accompanies a downturn in the business cycle. High unemployment always turns the labour market into a buyer's market. But more complex and profound changes are also happening. The postwar boom that was peaking when Lorraine Wilkinson first went to work at Champion Spark Plug resulted in increased real wages until 1976. Expectations and consumer spending rose along with the new sense of job security. After that a period of decline set in. In fact, from 1976 until the start of the depression of 1990, real wages shrunk by an average of 0.3 per cent annually.[20]

By the 1980s a new change was well under way. In that decade, fully 44 per cent of all new jobs added to the economy were in low-wage consumer and retail services.[21] What's more, many of these new jobs were in *precarious* employment or, as it is sometimes called, *non-standard* employment. Anyone who has looked for a job in recent years will be familiar with this type of work—part-time, temporary, or contract. It is work doing home sewing or providing home child care. It might be work keyboarding data into computers. It is work that is either part of the informal sector or done for labour contractors. In the 1980s, temporary-help agency work alone, mostly done by women, *tripled.*[22]

In non-standard work, as the Economic Council reported in *Good Jobs, Bad Jobs,* "Wage levels are generally well below those for full-time workers and fringe benefits are usually minimal." The Council's succinct, if arid, conclusion? "The implication of our research is that the labour market is offering economic security to fewer Canadians."

———————

"When we head into recovery we expect things to go back to normal and they didn't," says John Myles, a veteran analyst of the Canadian job market. "That's what's frightened and concerned people." Myles is describing the apparent boom that followed the Great Recession of the early 1980s. He does contract work for Statistics Canada, and he contributed to the Economic Council's final labour-market research project (just before the Mulroney Tories killed the Council). He now teaches sociology at Florida State University. A careful analyst who does not quickly jump to conclusions, Myles is a rare bird in an academic menagerie fixated on specialization: he is both an accomplished number-cruncher and a sophisticated theoretician. Until the late 1980s he was still cautiously optimistic, reluctant to join the growing number of observers who were looking with alarm at the emergence of an hourglass economy and a hollowing out of the middle of the labour market. But in the 1990s Myles became increasingly anxious about the trends he was seeing in his research, particularly with the phrase "jobless recovery" becoming commonplace.

"It's the distribution of working time that's really driving the polarization in Canadian earnings," he concludes. "Managers no longer manage their own workforce. What they're doing is managing contracts.... If we need three secretaries or software programmers we just order them up and keep them until we don't need them any longer."

The phrase is telling: "... until we don't need them any longer." With the ascendency of precarious, part-time work, and with competitiveness achieving the status of moral virtue, a brave new world of work is taking shape; human resources for sale in a latter-day war of each against all.

It is against this background that the training gospel has emerged. The most favoured solution to problems of insecurity and job loss hinges on convincing the workforce of the need to train and train again. Some call it lifelong learning. For others it is "learning a living." According to conventional wisdom, it will retool workers, giving them the flexibility they will need to compete in an emerging high-skill, high-wage economy. Everyone needs training to get those new jobs, to compete for them, to be able to get a grip on the wheel of technological fortune that is spinning ever faster. Politicians, business leaders, and the journalists who interview politicians and business leaders tell us about the new job requirements of "the innovation-based economy of the 1990s."

"It is clear that the performance of Canadians will be measured by a global standard of excellence," the country's largest business organization stated just as the training imperative was taking shape in 1989. "Most occupations are demanding higher skill and educational attainment or some form of vocational training.... In this environment, a highly skilled and adaptable labour force is essential."[23]

But despite all the talk of training, the reality has fallen short. Nowadays we're constantly hearing about people who have training, or university degrees, and still can't get jobs. There's a missing connection there, somewhere. As the chairman of the Business Council on National Issues told his audience of management enthusiasts, "Government labour market programs have failed to equip the unemployed with marketable skills." He made no mention of the private sector's woeful training record. For the past one hundred years, as we will see, Canada's manufacturers have consistently failed to invest in training their employees.

Small business, which ceaselessly claims to be *the* engine of job creation, has the worst record for training its workers. Big business does better, but invests its limited training dollars in managers and attitude adjustment programs for workers. Owners and managers in both sectors—and in governments intent on surrendering public services to business—benefit from a ready pool of people anxious and desperate for jobs. When the wheel of fortune spins, there are always more losers than winners.

There are disturbing indications that the growing polarization of the labour market is paralleled by a similar trend in training. For instance, a nationwide survey of firms conducted to assess the impact of computer-based technological change on workers revealed that the percentage of workers using computer-based technologies tripled between 1985 and 1991; and women were much more likely to work with these new technologies than men. The number of people trained for new technologies also tripled in the period 1986-91, a change accompanied by a shift in the occupational profile of computer trainees.

Whereas the preponderance of trainees in the early 1980s were [sic] clerical workers (55 percent), this pattern had changed radically by 1986-91, where only 21 percent were in clerical occupations. The loss of clerical training places was almost exactly matched by an increase in training of professional and technical employees in fields other than science and engineering, and of managers. Professional and technical workers, who do not even appear on the chart for 1980-85, accounted for just over one-quarter of those receiving technology-related training in the 1986-91 period. The share of training places occupied by managers rose from about one-quarter to about one-third during the same period.[24]

These findings backed up the trend underlined in the Economic Council's *Good Jobs, Bad Jobs* report, which came out foursquare for training as a key solution for both individual workers and the economy as a whole. Although most firms provide very little training at all, the training that is provided "is heavily concentrated among highly-skilled, well-educated male workers."[25]

The results of the 1993 National Training Survey (which gathered information from workers rather than firms) bear this out. The distribution of training was extremely uneven among the minority of workers who received employer-sponsored training. University graduates and those earning over $35,000 a year were three times more likely to be trained on the job than high-school dropouts or those making less than $20,000.[26]

A similar trend has been observed in the off-the-job adult education courses, where participation is also inequitable. As with the old aphorism about the rich getting richer, the educated get more education. Between 1986 and 1990 the number of Ontario adults taking continuing education increased from 20 to 31 per cent. But nearly half of all university graduates took courses, whereas only 5 per cent of school dropouts participated. Less than 5 per cent of adults were in job retraining programs, and 1 per cent were in basic literacy or English as a Second Language (ESL) programs. "There are indications here," concluded David Livingstone of the Ontario Institute for Studies in Education, "of a growing polarization of the class structure of our society into a 'permanent education culture' of permanently employed, highly educated core workers and, for the rest, a culture of temporary jobs with spasmodic bits of retraining and general upgrading."[27]

Just as some combination of skills, training, and education becomes the key to personal prosperity, postsecondary education is getting more and more expensive. Access to educational opportunity will increasingly be determined by the luck of the draw at birth. As Robert Reich pointed out, people "fortunate" enough to have had good education will extend this advantage to their children by "investing" in tomorrow's human resources. Good fortune and luck become ever more important in a gambler's society that looks at first blush like a meritocracy, sorting people out according to skill and ability. Those who don't make the education/training grade are consigned to the bottom of the hourglass, the basement of the split-level job market. It is no one's fault but their own—or at least this is the unspoken message, particularly in training sessions that emphasize individual skill resumés.

At a Kingston meeting called by the local Member of Parliament, the parliamentary assistant to human resources minister Lloyd Axworthy explained that although his Italian immigrant father had worked at construction all his life, he fully expected to see his children change "careers" a dozen times or so over the course of their lives. Maurizio Bevilacqua apparently believes that all of this is inevitable. In a world ruled by the market and the unaccountable corporations that dominate it, workers are so many empty bottles that have to be gathered up and rewashed and recycled continually, with many of them being broken up or rejected as inadequate along the way. These people are what historian David Noble calls "the human debris of progress."[28]

The limited protection that these people used to get is withering away. A dual process is unfolding on both sides of the border, with the United States, as usual, in the lead. As the labour market is transformed, with the middle being eroded and the bottom expanded in both countries, welfare state protection for the people at the bottom has been slowly dismantled, although due to the strength of the labour movement and social-democratic parties the task has been made longer and more difficult in Canada. The welfare state is, after all, better developed in Canada than in the United States. But the well-documented fact remains: social programs offer less and less protection to the people who need them most.[29] When Bevilacqua came to Kingston, it was as part of a national effort to sell a further dismantling of the welfare state by lowering UI benefits and raising the costs of education.

This is perhaps the end result of the society that C. Wright Mills described in 1951 as "a great salesroom, an enormous file, an incorporated brain—a new universe of management and manipulation."[30] The hard-boiled American sociologist worried about people being dependent on large organizations, on jobs in which they were simply told what to do. Politics would be reduced to the level of spectator sport, with people losing a sense of their ability to be active participants in a struggle for the public good. But at the time Mills was writing, such political-cultural issues seemed remote to most workers, whether blue collar or white. The age of rising expectations was in full swing. In postwar Windsor, for instance, they were busy producing and promoting North America's most cherished symbol of progress.

Orwell also imagined it all when George Bowling entered an early version of a plasticized McDonald's-style eatery where a young woman with an early version of a McJob served him some "phantom food." With "a sort of propaganda floating round . . . to the effect that food doesn't matter," Bowling dutifully bit into his rubber-skinned frankfurter.

> The thing burst in my mouth like a rotten pear. A sort of horrible soft stuff was oozing all over my tongue. But the taste! . . . I got up and walked straight out without touching my coffee. God knows what it might have tasted of. . . . It gave me the feeling that I'd bitten into the modern world and discovered what it was really made of.

That was in 1939. *1984* it's not; 1995—and beyond—it is.

2

WINDSOR

"You Could Quit at Ford's in the Morning. . ."

ON THE TABLE beside Dari Desroches's fridge sits a computer printer, its corner propped up by an empty can. The can's label is stained with beet juice, but you can still see the familiar features of the Jolly Green Giant smiling out at you across the room.

The contrast in the place is jarring. Crowded onto the table beside the computer are neat piles of papers, cancelled cheques, and print-outs. There are more print-outs on the floor, spreading out into the living room a few feet away. "Each one of those is a company," Dari Desroches says, pointing to the print-outs. She is a single mother supporting herself and two young teenagers, and the room in South Windsor, Ontario, also contains the more conventional clutter of a busy family kitchen: kids' lunch boxes, coffee cups, dishes, tea towels. A soccer ball on the floor seems to account for the broken window that looks out onto a pleasant, tree-lined street. The road itself looks new. Desroches tells me it has been freshly paved following the installation of new sewers.

"Last year I had to hook my sewers up. It cost fifty-eight hundred dollars," she says. "You have to connect your sewer to the city's new line. It's not bad enough that they raise your taxes six hundred dollars." Unable to pay for the whole thing in cash, Desroches ended up doing bookkeeping—"on the never-ever plan"—for the small firm that supplied the pipe for the job. Desroches assumes she'll be doing the books until her debt is paid off.

Dari Desroches is a tall, blonde woman with an easy-going manner. She grew up in Windsor, where her father had a part interest in a small heat-treating plant that produced hardened metal parts for the Big Three automakers (Ford, General Motors, and Chrysler). She worked in her dad's company for seventeen years, mainly in the office, but she would also go out on the shop floor when the need arose. The company's three huge customers could dictate the terms and conditions of supply, and the result was a roller-coaster business with lots of downtime and plenty of overtime. Still, the unionized workers in the plant could gross between $36,000 and $40,000 a year if they worked for six or seven days a week.

Sometimes Desroches worked at sorting screws until her fingertips were numb. "I worked the belt lines when it was 130 degrees and Dad couldn't get anyone else to work Sunday. The parts had to get out or a GM line would shut down."

Most of the time she did purchasing, shipping, receiving, and bookkeeping. Then, when computers arrived in the plant office, she found it only took her a day to close out the month's accounts, a task that used to take a week. Foreseeing that the company was headed for a one-person office, she decided to learn more about computers. She bought one for herself and began courses at the University of Windsor to work towards a certificate in general accountancy. But in 1992 the company—along with dozens of other small plants in Windsor—disappeared altogether as a combination of recession, free trade, and corporate restructuring devastated the city's auto-parts sector.

Desroches has now joined a growing group of Windsorites. In the jargon of the labour-market specialists, she is part of a "contingent" workforce relying on "non-traditional" work. Most of the people she knows are doing work "under the table." At first, after her regular job ended, she told herself that she was going to find a full-time job. Instead she found herself drifting into the realm of self-employment.

She describes herself as a freelance bookkeeper. In the winter she does spreadsheet work for a trucking and snow removal business. The drainage company has work for her from March to November. She does books for a blacksmith and a trainer at the local raceway, a couple of small restaurants, a travel agency, a personal counselling service, and a dozen others. Some of the clients bring the books to her house. More often she uses her aging Oldsmobile for pickup and delivery on a monthly or quarterly basis.

Like many women she knows, she has mixed feelings about her new line of work. For one thing, the job allows her to be there for her kids. She can design her own hours. "As a single parent I like it. I can get my kids out the door and onto the school bus in the morning and not have to punch a clock. I can go on their class excursions. I can be there anytime that they need me." She says "being a family is more important than any work."

Desroches is aware of how a sense of insecurity has hurt her family. She often feels tired—not surprising given her mix of family care, multi-freelance jobs, and one night a week at the University of Windsor in the certified accountancy course. But in addition she waits on tables at lunch hours for one of her restaurant customers. She has filled in as a bartender for another. "You have to be able to do so *much*. I have made sure that I am flexible. I can do just about anything."

Unlike some of the displaced workers she has met at government-sponsored adjustment courses, she is not afraid of computers. But she knows that the young people graduating from the high school her children attend are familiar with computers. They will soon be competing with her.

"They're asking, 'What do we do to secure our future?' They see us hustling and taking what we can get when we can get it. I've got a technical knack," she admits. "I can take the top off my computer. I switch hard drives. I just take a screwdriver to it."

Her self-confidence masks an anxiety that is almost palpable. It is clearly not easy to live with her scattered work life, with the computer in the kitchen and the files seemingly everywhere else. Most kids look forward to Christmas and the day that school finally finishes at the end of June. Desroches's children also look forward to the day after April 30 and the passing of Revenue Canada's tax deadline, when the piles of paper shrink and everyone can relax, for a while at least.

"This is about the largest house I can afford," she admits, looking around her bungalow. In such a small place it is impossible to escape the work, and the pressures. "It's all around you, all the time."

When the children notice that their mother is not out and about with her usual zeal, they become edgy and start looking through the newspaper's help-wanted section and pointing out possibilities. "Mom, are you going to put in for any jobs as a bartender?" "Did you find anything today, mom?" "When's your next interview?"

Windsor calls itself "The Border City"—witness the long stretches of strip development cluttered with signs like "Border City Collision" or "Border City Auto Glass." Canada's southernmost city sprang up along the south shore of the Detroit River from white settlements that took root after a French pirate and fur trader named Antoine Laumet de Cadillac planted the French flag on the opposite bank in 1701. Its subsequent growth as an industrial town derives from the happy accident of its location across from what would become the American industrial heartland. The connections between the cities are so strong that when the first elections were held in the newly designated Upper Canada, several people from the U.S. side were elected to the British colony's House of Assembly.

Most people associate the name of Motown's founder not with an eighteenth-century carpetbagger but with an expensive set of wheels, but although the Windsor-Detroit area is synonymous with the auto industry it also has a rich agricultural hinterland. Hiram Walker, a Detroit grocer's clerk, used grain from the surrounding region when he set up his distillery on the Canadian side in 1859 and started to produce quality whiskies, including the Canadian Club brand. A salt formation in the shape of a half-grapefruit underlies the area, and it spawned salt mines and early pharmaceutical industries.

A few years after the turn of the century Henry Ford began to produce cars in Ford City, just east of Walkerville. The town grew in lockstep with the expansion of an industry and a culture fully submitted to the tyranny of internal combustion. Business assessments rose from just under $700,000 in 1907 to almost $5 million in 1927. The population, 28,064 in 1918, rose to 47,177 in 1923 and then to 66,893 in 1927. Most of the new residents came to get jobs in the car plants that crowded around "the works" built by Ford.[1] The population swelled yet again in the early 1930s with waves of Eastern European immigrants pulling back from desolated prairie homesteads. Entire Ukrainian and Polish settlements uprooted themselves to head east for industrial Ontario. By then the Big Three as well as a crowd of other now-forgotten auto companies had developed strong internal labour markets, with jobs going to families and friends of workers. Many of the recent arrivals had to purchase jobs using church and ethnic connections.

Windsor is unique among Canadian cities not only because it looks *north* across the border to the United States, but also because it is alone in mirroring Canada's relationship with the United States— it literally sits in the shadow of a great American metropolis. On a clear winter night you can look across the narrow Detroit River at Detroit's skyline, the Renaissance Centre, the Joe Louis Arena, and the other modern fortresses that hide the fires of the homeless who inhabit the decaying metropolis beyond. On a hot evening in July 1967, Windsorites crowded their side of the river to watch Detroit burn as the black poor and Lyndon Johnson's National Guard fought it out with Molotov cocktails and rifles. Detroit became Berlin, 1946, a ghost deserted by all but an underclass who could not afford to move away. "The city many Windsorites regard as an extension of their own," wrote the first Canadian reporter on the scene in a masterful bit of understatement, "will somehow never be the same."[2]

The great boxer Joe Louis, a Detroit resident, had worked in Henry Ford's U.S. factories and in the time of the Great Depression became a symbol of the little guy fighting against the odds, the first black heavyweight to make it to the top. Detroit and its smaller Canadian cousin have been typical of the success of North American manufacturing for most of this century, riding an unequalled tide of prosperity. There were jobs for all, the Poles of Hamtramck, the Serbs who lived in the area around Drouillard Road in Ford City, the town that became East Windsor when the first Chrysler rolled off the line there. Until the 1967 riots, half of Windsor would flock over to the Detroit side to spend their paycheques at Hudson's and the other big stores in downtown Detroit; in fact, Windsor has never been able to attract a major department store to its forlorn downtown.

The cities by the river were a wellspring of a society into which people could apparently be smoothly incorporated as part of the great consumer middle. Those who confuse class with status saw this as an expansion of the middle class. But just about everyone, it seemed, had a share in the mass market of goods, especially cars. Working-class jobs with good pay were widely available. The white-collar middle class prospered. General Motors seemed infallible. A revolution of rising expectations would roll over past uncertainties. The crest on the front of the Detroit Red Wings sweater, after all, was a wheel with wings.

Of course, the road to the apparent prosperity of the 1950s had not been smooth. The war years were crucial. The dark days of the Great Depression were over. With production humming and the Big Three turning out machine guns and tanks, there was suddenly a labour shortage. Women were even allowed into the Windsor plants. Workers gained strength as the labour market became a seller's market; the autoworkers and their union began to expand the initial gains they had made in Windsor in the 1930s, when they had taken a cue from their bosses. The car companies had already moved across from Detroit to use Windsor as a bridgehead in their first moves to globalization, taking advantage of Canadian production to gain access to Imperial Preferences and expanding markets in Australia and South Africa; Ford of Canada became the Empire's biggest automaker. The U.S.-based United Automobile Workers also crossed the river to gain a toehold in Canada, organizing their very first Canadian local in Windsor, where the workers at the Kelsey Wheel Company established Local 195 in 1936.

The Windsor autoworkers struck the Ford Motor Company nine years later, marking a turning point in Canadian labour history. The war had enabled the unions to make solid organizing gains, and the atmosphere on the job had changed since the early thirties, when the employers held the whip hand.

"You think you're pretty goddamned smart," growled one boss. "Someday this war's over and we'll fix you bastards then."

"Partner, on the contrary," sneered the worker confidently, "Hitler's not going to win this war."[3]

This sort of working-class assertiveness gave rise to the Ford strike of 1945. Mackenzie King's ruling Liberals, already worried about a repeat of the general strikes that rocked the land after World War I, and spooked by the growing electoral strength of the Co-operative Commonwealth Federation (CCF), were prepared to grant concessions to the left both on the job and in the area of social policy. When the autoworkers walked out at Ford Windsor for three months in late 1945, the result was a giant step forward for the trade union movement in terms of stability of organization and the smooth flow of collective bargaining. The "Rand Formula," which took its name from the arbitrator of the Ford Strike, gave the union the right to collect dues from all workers in any plant it organized, for the simple reason that all the workers benefited from collective bargaining. With this assurance of financial stability the unions could become more institutionalized in their operations.

The new deal was not without its benefits for the big corporations. They gained stability on the job, and could rely on the unions to police their hard-won contracts and ensure that the Plymouth and Ford sedans flowed smoothly off the assembly-line. As Liberal

MP and former Windsor mayor David Croll said at the time, "What Mr. Rand has done is obvious. He has taken management-labour disputes out of the brick-and-tear-gas stage [and] delivered potential members to the union and given it union security. And to the company he has given a measure of security as well, protection from wildcat strikes."[4] It all added up to a move away from what historian Eric Hobsbawm has called collective bargaining by riot.

The Ford strike was also important locally, for it illustrated the extent of working-class power in Windsor. After the men of the Windsor police force refused to move against the autoworkers' picket line with the zeal expected by the authorities, the local Police Commission—against the wishes of the mayor—wired the provincial authorities and requested reinforcements. The workers responded, appropriately, with a barricade of motor cars erected in a T-formation on Drouillard Road and Sandwich Street. To this day it is said that anyone wanting to occupy the mayor's chair at Windsor City Hall must first secure the support of the Labour Council; local companies that face job action attempt strikebreaking at their own peril.

The car industry around Windsor became a laboratory for the historic postwar compromise often referred to as "the deal." Under this accommodation, unions obtained security for thousands of full-time jobs, seniority rights, and wage stability.

"The keystone of the whole system," wrote Michael Piore and Charles Sabel in an influential study, "was the wage-setting formula negotiated between General Motors and the United Auto Workers in 1948." The new formula linked increases in labour productivity to the consumer price index. "Wages, it was agreed, should rise by this amount every year. . . . Private-consumer purchasing power would expand."[5] Businessmen, who had already joined the pioneering Henry Ford to embrace advertising and the gospel of consumption, had no objection to the further expansion of a mass market in which workers would be free to choose from the cornucopia offered by the goods society.

Implicit in the deal was the idea that workers would have little or no choice but to accept the control of the boss over crucial questions of how their skills were deployed, over technological change, and over other management "rights." Similarly, labour turned away from its historical demand for shorter hours without a cut in pay: for a century the workers' movement had agitated for a new way of thinking about what constituted a "normal" workday. After the war the work-week got stuck at forty hours to the extent that—a half-century later—we still think of forty hours as the benchmark. In an essay entitled "The End of Shorter Hours" historian Benjamin Hunnicutt concludes, "The concept of free time as leisure—a natural part of economic advance and a foil to materialistic values—was abandoned."[6] Unions bowed to prevailing notions about how the winged wheel of progress would keep spinning on and on, indefinitely. It was all part of an orthodox tapestry stitched together by a faith in endless consumption, continuous growth, and lots of time on the job for lots of men. Women were expected to work at home, as "homemakers."

For those born in the Windsor area during and after the war, the fifties, sixties, and to some extent the seventies were the years when jobs were easy to come by. As they say over and over again in Windsor, you could "quit at Ford's in the morning and get hired at Chrysler's in the afternoon."

Russ Jackson is typical of his generation. His father commuted every day to work in Ford's Windsor foundry. Born in nearby Chatham in 1959, Russ married Brenda Ritchie, the girl next door, after teaching her how to drive the family car. At age eighteen he followed his father and started work at the Ford foundry, then watched as his parents and their friends retired and began spending half the year in winter homes purchased in Florida. Rather than waiting for the slump of 1979 to end, Jackson turned his back on the seniority list at Ford because he wanted to avoid the Unemployment Insurance runaround. He took a job at a newly opened foundry that supplied forged engine parts to the Big Three.

Working at Brant Casting was filthy, dangerous work, and Jackson spent his days sucking in silica dust and smoke. He began calling in provincial labour ministry inspectors, and they eventually closed the operation until it was cleaned up. In spite of the hazards, he recalls the early days at the foundry with fondness. Most of the other workers were like himself, men in their twenties or early thirties, men just starting families in what he calls a "kid boom."

Jackson is a burly, bearded bear of a man whose gentle nature seems incongruous at first. He and Brenda and their three girls live in a subdivision called Forest Glade, not far from a main intersection where the Tecumseh Mall and the Eastown Mall compete for passing traffic. Aside from Windsor's mild climate, the scene could be anywhere in suburban Canada. "You look back at what your parents had at this age and tell yourself, 'I want to do this and that.'" His parents, he says, "were able to buy things." His father had worked seven days a week, whether he wanted it or not, for as long as Jackson can remember.

That older kind of job security now seems like an accident of history to Russ Jackson. Ten years after he signed on at Brant casting, Jackson and his mates punched out for the last time when the plant closed for good, one of the first of many factories to feel the first bite of the recession that began in 1990. For ten years Jackson had been living from cheque to cheque, saving little. By then Brenda was working as a moulder in a plastic factory run by the Complax company. She had started back to work when she felt her youngest daughter was ready to be left with a caregiver. Less than a half a year later the Complax plant, though new, also closed down.

A day before it is due to close I tour the nearly deserted plant with Brenda Jackson. We watch as a few of the remaining workers etch their initials in the wet cement poured to fill the holes where the giant presses had stamped out plastic parts for cars.

"There is no other plant in Canada or in the U.S. with the high technology we have," company president Ralph Zarboni had bragged when the plant first opened in the 1980s.[7] "It's state of the art," Brenda Jackson says. She is still proud of her former workplace.*

"The days of getting hired somewhere and spending thirty years there are done," Russ Jackson says. "Now you'll be jumping from job to job. You can feel it in the town, the insecure feeling that most people have. It's affected our kids." He tells how his daughter wanted to go ice skating but couldn't because her skates didn't fit. "She didn't want to ask for the ten bucks for another pair."

In ten years at Brant Castings Russ Jackson managed to put $600 into an RRSP, and now he can't rely on getting a pension. "People who jump from job to job don't have that. That's the scary part. We're all going to get older. How will we live then?" The very notion of retirement—to say nothing of the "early" retirement now aimed rather desperately at saving the jobs of younger employees—was only being invented when Jackson's dad started work.

For much of this century business managers have applauded their apparent success in shifting the social relations of production by breaking down their employees' tasks into smaller and smaller parts that can be repeated over and over again. This shifting of production puts more control into the hands of engineers and other professionals. The process is usually called "scientific management," and there is much talk of the struggle for control over the "labour process." Autoworker Ben Hamper, who toiled on a GM truck line in Flint, Michigan, called his work "repetition as strangulation" in "a huge metallic ant farm."[8]

---

* Complax closed its Windsor plant as part of a complicated reorganization of the troubled Edper Bronfman empire. Although the auto-parts maker was in good shape compared to Edper's many other holdings, profitability was not enough to save the Windsor operation from being caught up in some much larger corporate machinations. The Complax plant eventually changed hands as part of a deal that saw Chrysler receiving government money to set up a training centre there.

Whatever the label, the postwar work pattern in industrial cities like Flint and Windsor was clear. Henry Ford liked to boast that he could train a worker in two weeks. One visitor to a Ford plant in the early years marvelled at how the assembly-line didn't just turn out "flivvers" (small, cheap cars) but also turned workers into the extensions of the machines they operated. "Every employee seemed to be restricted to a well-defined jerk, twist, spasm or quiver resulting in a flivver," he wrote in the *Tri-City Labour Review*. "I looked constantly for the wire or belt concealed about their bodies which kept them in motion with such clock-like precision." The writer said he would never again "be able to look another Tin Lizzy in the face without shuddering at the memory of Henry's manikins."9

Many routine assembly-line jobs needed little in the way of what is commonly seen as skill, and that was the way the corporations involved wanted it. Still, tool-makers and electricians closely guarded their crafts and the power that went with them, and in the context of the big industrial unions they shared that power with their less skilled co-workers. This on-the-job strength extended to the labour market at large as long as traditional working-class jobs were widely available; "You could quit at Ford's in the morning . . ."

By the 1990s supposedly "unskilled" but relatively well-paid workers laid off from unionized manufacturing jobs in plants like Complax and Brant found themselves scrutinizing want-ad pages offering jobs as part-time telemarketers working from home. It had become noticeably more difficult to get by on hard work and merit alone.

Brenda and Russ Jackson are certainly no strangers to hard work. At Complax Brenda worked "midnights"—the graveyard shift that ended in time for her to get home to see Christie and Jamie, her two older daughters, off to school. Then she would look after Erin, still a toddler, until nap time in the early afternoon. After school the two older girls could look after their younger sister and Brenda would sleep until Russ came home from work.

Brenda Jackson had taken a job at a tool and welding plant after Christie was born. Russ Jackson had just been laid off at Ford. She stayed at the plant for two or three months, until her husband went back to work and wanted her to quit. "My wife will never have to work," Russ used to say. "At that time you didn't need two incomes," Brenda says. Now, she says, Russ has changed his tune.

Brenda Jackson keeps her daughters on a short leash, refusing to let thirteen-year-old Christie go to the mall, no matter what her friends are doing. "I'm not stupid. I used to meet your daddy at the mall." Brenda's nickname at Complax was "Sybil," a reference to her ability to switch from an apparently contented press operator to the screaming scourge of the supervisors, just like the character "Sybil" in the book about multiple personality disorders. Like her nineteenth-century forebears, the iron moulders who produced some of the most tough-minded unionism in the early days of industrial Canada, this plastic moulder has both a suspicion of authority and pride in her work.

Although her job was typical of work described as unskilled, Brenda Jackson figures it took her a year to learn how to operate the press properly. She evaluated, cut, and weighed the plastic before wrestling sixty-pound batches into place. "You gotta be experienced. It was a hard job; you screw up one button and you're done." Touring the Complax operation we catch sight of the one remaining press not yet shipped out. Its computerized controls are inset into a sealed box, its monitor still blinking "SMC Press Control System prepared for John T Hepburn Ltd, Toronto for Complax, Dec 8, 1986 Version 1.34."

"You've gotta learn the technology of their darn computers," Brenda Jackson says as we look at the screen. "On steady midnights we lucked out because we had a foreman who taught us a bit about it."

The jobs at Complax were worth twelve dollars and change to Jackson and the other workers at her wage level—four out of five of them women, many recent arrivals to Canada. Brenda Jackson is relatively lucky in that English or literacy pose no barrier to her. Still, two years after Complax closed she found herself on the training treadmill. She went back to school and got the Grade 12 qualifications necessary to get into a computer programming course. Like count-less other Canadians, she regarded familiarity with computers as im-portant—and enrolment in a course vital to maintaining her UI claim. But by 1995 Brenda Jackson was still waiting to get into the programming class. Demand for spaces was high, and funds were limited.

Russ Jackson did find a secure job after the foundry closed. He parlayed his experience as a union health and safety activist into a job as a caretaker at a housing co-operative sponsored by the Canadian Auto Workers. In a labour town like Windsor, the union culture offers workers options they do not enjoy in places where the move-ment is weaker. Still, Jackson is far from complacent about the prospects for people like himself.

"Our only hope is to get into retraining or something," Russ Jackson says. He figures their kids will have to move away from Windsor and go to a bigger city to get the education they will need. "It's scary out there. The jobs that *are* there need higher skill. There'll come a point where some of our generation or our kids will get those jobs, but they'll need to be a lot more educated, more skilled." The companies, he says, will have "all the robots they want, but the robots don't buy cars or groceries or houses." The companies "can jump to Mexico for two dollars a day but that's not helping Canada any."

The former foundry worker would probably feel right at home with the theory in Antonio Gramsci's essay "Americanism and Fordism." In the 1930s the Italian social theorist speculated about "the deal" even before it was formalized. Gramsci was writing from a jail cell to which he had been confined for communist activity by Mussolini's fascists. Given time to think, he identified the tendency of workers in the United States to lose control over the labour pro-cess but gain higher wages.[10] Henry Ford had it figured out while Gramsci was still a student in Sardinia: he outraged his fellow busi-ness magnates when he unilaterally raised the pay of his Detroit workers to five dollars a day (four dollars across the river in Wind-sor). Ford knew that if they were to be able to purchase commodities en masse—including his Tin Lizzies—the workers would need to have something to purchase them with, either cash or credit.

Brenda and Russ Jackson each drive mid-size American cars that are starting to show their age. The couple personifies the fears of the "retail analysts" so often quoted by journalists chasing comment about the stubborn nature of the 1989-93 recession and the unwillingness of consumers to go out and buy those big-ticket items. The usual label for this faltering allegiance to the temptations of the market is "lack of consumer confidence." The Jacksons don't have much consumer confidence, for the simple reason that they don't have much confidence in the labour market. They are ultra-careful about their purchases, clipping coupons for supermarket specials.

In Windsor between 1989 and 1993 fifty plants, including the one that Russ Jackson worked at, closed as recession and free trade hit hard. According to local labour-market analysts at the Canada Employment Centre (CEC), 4,300 manufacturing jobs vanished. Short but frequent mass layoffs of between 800 and 8,000 workers hit the auto-assembly sector. Overall unemployment rose from an already high 8.1 per cent in 1989 (*before* the recession) to 12.8 per cent in 1992.[11] By the fall of 1993 there was some good news when Chrysler finally agreed to the demands of the Canadian Auto Workers (CAW) to bring in a third shift at the minivan plant, creating a thousand new, well-paid jobs. As usual, lineups snaked around area CECs as thousands waited to get application forms. But the 1989-93 downturn was unlike the normal periodic slumps in auto that historically would end as demand picked up. Many of the feeder plants like Kelsey-Hayes, where the Canadian autoworkers first organized successfully, went south for good. The wheel and rim maker had been one of the top ten manufacturing employers for decades, employing 700 people even at the height of the recession of 1981 and bouncing back to 835 by 1986. Also gone (to Burlington, Iowa) never to return was Champion Spark Plug (435 workers in 1981). Wyeth Pharmaceuticals (235 workers—mainly women—in 1981) closed in 1993 after promising job stability if only the Mulroney Tories granted extended patent protection to drug multinationals. Wyeth took millions in government subsidies and departed to centralize its operations in suburban Montreal.

The list goes on down into the smaller feeder plants (known locally as "finger factories" in reference to their safety records) like Brenda Jackson's "high-tech" Complax Plastics, firms that supplied GM, Ford, and Chrysler. Compounding this trend in manufacturing job loss was a dramatic shift in employment patterns among the Big Three. In 1993 Chrysler announced a massive retooling and expansion at the minivan plant: "$600-million buys no auto jobs" said the main headline in *The Report on Business* the next day. Ontario's NDP kicked in $30 million for training. GM's new Windsor transmission plant would mean five hundred fewer jobs.[12]

"There's no question that this massive new manufacturing investment is not generating the kinds of big job numbers that were present in the sixties and seventies," says Paul Bondy, head of the Windsor-Essex County Development Commission. Bondy's ancestors came to the region as part of a wave of eighteenth-century French immigration. The long, narrow French-Canadian farm lots gave rise to the long blocks that reach back from the Detroit River, streets with names like Ouellette (the main drag) and Pierre (now pronounced "Peary") that bespeak the city's francophone origins in the same way that Chrysler and Cadillac streets reflect its recent industrial prosperity.

Bondy's job is to help shore up that prosperity by attracting new industry to Windsor, one of Ontario's half-dozen—but Canada's few—true manufacturing locales. The front of his Commission's glossy handout is emblazoned with the word "SUPERMARKET" sitting on a row of eleven flags that apparently signal the global economy. The Canadian flag is in the middle; the Stars and Stripes is prominent at one end. No one in Windsor has a better grasp of local economic trends than Paul Bondy. "Even as late as 1978, when Ford built the Essex Engine plant, $768 million was the equivalent of

2,350 new jobs. Ford is now spending $1.5 billion to retool a plant to manufacture a new generation of truck engines. Basically they're going to *protect* 2,000 jobs."

Bondy's job calls for a certain optimism, but a good deal of his hope stems from Windsor's geographic advantage. He denies that free trade has had a major impact. The city is within an hour of all the engineering brains of the Big Three, and although GM's successful Saturn line may be bolted together in Tennessee, he emphasizes that Saturn's brains are "here in North Detroit." And "A lot of us don't even consider the border as a border," he says. For Bondy, the future of Windsor is the future of the auto industry; he believes the proximity to the core of the midwestern U.S. automotive industry will see it through. He is probably right. As long as the Canadian dollar remains low enough, the city will not shrivel completely. There will be some good manufacturing jobs. The boarded-up storefronts downtown and the job lineups could be worse.

"Maybe people don't look far enough afield and realize what's going on in Oshawa and St. Catharines," he says in reference to other traditional manufacturing centres suffering more than Windsor. "If you can hang on to what you've *got* these days you're doing really well."

Still, the cover of Paul Bondy's "SUPERMARKET" directory provides clues to deeper problems. It contains crisp photos of dozens of products from the Windsor area: Windsor salt, Heinz ketchup from nearby Leamington, a Champion spark plug (it will have to go), crankshafts, transmissions, and other auto parts, a bottle of Canadian Club whiskey from Hiram Walker's old riverside distillery, a pile of produce from the rich surrounding farmland, a Cyclette dispenser with a month's supply of Wyeth's Triphasil birth control pills (this will also have to go), and—dominating the layout—a Chrysler Magic Wagon minivan. But what of the shift from manufactures to the less tangible products of the expanding service industries? If it has not already, the Windsor Development Commission will now have to add another image to its industrial bundle: the roulette wheel.

The ketchup bottle, temporarily threatened, will remain. The Heinz corporation has been making ketchup in nearby Leamington since 1910, employing fourteen hundred people as recently as 1988. By 1994 there were eight hundred jobs left. With the coming of free trade the company was well positioned to bludgeon its workers into accepting lower wages (they had been making $17 an hour). When the workers rejected two offers that had been accepted by their United Food and Commercial Workers negotiating committee, Heinz's first-ever North American human resources vice-president sent layoff notices to 450 workers, some with twenty years' service. The workers gave in and accepted concessions, Leamington handed Heinz a $150,000 tax break, and Vice-President Les Dakens summed it all up: "Free Trade created a number of jobs like mine."[13]

These days working-class people in Windsor tend to repeat the words *scared*, *afraid*, or *insecure* in conversation. It is now common for people in their forties to have sons and daughters—and often grandchildren—sharing the family home while they take another retraining course and/or work delivering pizzas. A host of complex trends are at work here. Money moves back and forth across the globe at a breakneck speed. With capitalist enterprises spread around the world, steel, electronic equipment, and automobiles can be produced anywhere from Tennessee to Taiwan. A new spatial division of labour has loosened the ties of an industry like auto to any particular region.

Using computer technologies, corporations can produce more goods with fewer workers; between 1975 and 1990 an increase of 5 per cent in time spent processing material translated into a 26 per cent surge in the real value of production.[14] The core of working-class jobs in industrial towns like Windsor becomes smaller as the occupational structure is increasingly fragmented into various parts of a now-dominant service economy. As the Jackson family found out, the last generation has followed their parents through the plant gates.

By 1992 the wheels were apparently falling off Windsor's reputation as not only a working-class town but also a *prosperous* working-class town. That spring a local columnist tried to boost flagging civic spirits, chiding Windsorites for their "monumental inferiority complex." Even though the city had a bad air-pollution problem, high rates of respiratory illness and cancer, and a "devastated" downtown, wrote Gord Henderson in the *Windsor Star*, fellow citizens should take pride in being one of Canada's "leading industrial centres." According to Henderson, "This tenacious city doesn't just shuffle bits of paper for a living. It earns its keep with skill and sweat and needed products from mini-vans to precision machinery."[15]

Six months later the province announced its choice of Windsor as the site for Canada's most extravagant casino. The operation would create eight thousand jobs, mostly in hotels and restaurants—the "hospitality industry." When the news broke, merchants rushed from their Ouellette Avenue shops to clap Mayor Mike Hurst on the back. Enthusiasts talked about how Windsor's strategic location had once again come through; upwards of twenty-five million people lived within a four-hour drive of the new casino. The Labour Council was solidly behind the project, and the chorus of cheers easily drowned out dissenting voices warning of the crime and crowding that casino gambling might bring to town.

Tourism has always represented an elusive dream of diversification for Windsor, a gritty place that people have usually passed through on their way to somewhere else. "Gambling," says optimistic restaurateur Kurt Deeg, "is a destination attraction."

Deeg has been in the business for twenty-seven years, running a pair of carriage-trade establishments on downtown Chatham Street. He spent weeks personally collecting twelve-hundred signatures to attract what he calls "the gambling industry." As far as the declining auto-parts industry and stagnating job prospects with the Big Three are concerned, Deeg simply shrugs: "If Canada wants to survive as a force in this world, it's a fact of life. This is a world revolution. There's nothing to do but hope our workforce is—and I think they are—better trained, have a better attitude, have a sense of pride in order to preserve that job."

"We're hot. We're big-time hot," glowed Mayor Hurst, predicting a boom in tourism-related development: a marina, an aquarium, a convention centre, a sports complex, new hotels. "It's a job-creation opportunity the likes of which we haven't seen for years and years."[16]

In 1993, to take the pulse of the labour market, the federal government's local Canada Employment Centre together with the Windsor-Essex Skills Training Advisory Committee commissioned a detailed survey of 345 firms. They found that most manufacturing jobs were full-time and there were still many good jobs in skilled trades (43 per cent of occupations reported were skilled trades). Traditional personal and retail services reported two out of three jobs as part-time. But the lowest overall job growth rate was in Windsor's key sector, the motor vehicle and motor vehicle-related manufacturing industry. "Higher growth rates in every other industry" were predicted. The study agreed with Development Commissioner Paul Bondy that the best Windsor could do in auto was to hold its own: "The objective is not to 'win the most,' but rather to 'lose the fewest' automotive businesses and manufacturing jobs."

Like many such studies, this one could only recommend diversification as the key for a city wanting to ensure its citizens access to decent jobs. It dismissed the gambling and tourism initiative as at best a temporary labour-market fix.

Gambling jobs, it said, tend to "add little value, require minimal skills, command the lowest wages, and are largely part-time rather than full-time positions." The jobs, though needed, would probably not pay good wages or provide the labour-hours necessary to counterbalance the full impact of lost manufacturing jobs. "The casino will not restore prosperity to the average citizen in the area; for many though it may represent the opportunity to survive while looking or preparing for an occupation in another field."[17]

The Windsor casino was an immense success from the start, providing its U.S. operators with big profits and the Ontario government with a tax windfall. Wages for casino workers averaged $8.75 an hour, although some of them got little more than the minimum wage. The fact that Windsor is such a strong union town, however, led to quick affiliation with the CAW. In 1995 the low-wage service-sector workers at the casino did not hesitate to stage a strike, which turned out to be successful—something unheard of for the millions of other such workers across the country.

As in Windsor, so in Canada as a whole the postwar accommodation between capital and labour is clearly at an end. The "deal" is off. Stability is not part of the labour market; real wages are in decline. We are moving to a more inequitable, less fair division of wealth. Between 1973 and 1991 the richest 10 per cent of families increased their share of total income by 8.1 per cent. Until 1987 the people in the middle 20 per cent of families held their own. But in the period since then (when Dari Desroches and the Jacksons lost their jobs), this middle group began to lose its share. By 1991 the bottom 60 per cent of families with children had a smaller piece of the pie than they had in 1973—a change affecting all families who earned less than $46,578 per year.[18]

The debate over the shrinking middle started in the United States in the early 1980s when academic observers noticed the hollowing out of the blue-collar middle in the area around Boston. The same pattern has repeated itself across the river from Windsor, where, according to State Representative David Hollister, chair of the Michigan legislature's Social Services Appropriations Committee, 70 per cent of the new jobs pay $7,000 or less. Hollister calls it a "dysfunctional economy."[19]

Jim Brophy, director of Windsor's occupational health clinic, worked on the line at Chrysler before developing an interest in worker health and safety. He sees Windsor as "a fairly egalitarian community." But that is changing with the changing shape of jobs in the city. In Windsor, he says: "There's not enormous wealth and enormous poverty and all kinds of levels in between. By and large there aren't really ghettoized areas. Most people live in basically the same type of housing with the same level of income, mainly because the unions have been so strong and the wage levels have been good. That's been based on the auto industry and the fact that the vast majority of people here have worked in auto."

As the auto industry begins to "downsize" and as free trade makes it possible for the parts plants to leave, the working class is being differentiated, divided between the good jobs and the bad jobs sectors. With this division comes the possibility that Windsor's egalitarianism—and that of the nation as a whole—will be further eroded. Brophy says, "Yes, the people who are left in those big plants have fairly secure jobs and there's decent money for them. But there is going to be less and less of those people."

The growing number of working people at the bottom could very well begin to see those with good, unionized jobs as people with privileges that are not deserved. The workers with good jobs, uneasy because of what is going on around them, could just as easily come to regard everyone from UI recipients to underemployed women and minorities as a threat to their security. The changing world of work—which should be a place of opportunity—becomes a breeding ground for resentment, intolerance, and fear.

3

KINGSTON

"... With Not Much in Between"

A FTER her daughter Tiffany was born, Heather Dixon was not sure she wanted to get a job. She stayed home for two years before finally deciding that the family needed more than a single paycheque. One day, while shopping at the Biway on Princess Street, Kingston's main drag, she saw a help-wanted notice in the store window. She applied, and got the job. Because a regular day care would have swallowed seven out of every ten dollars she was clearing at the Biway, she arranged informal care for her daughter.

Soon after she started work Heather's unease about her decision turned into a full-blown case of guilt. Tiffany began behaving strangely, waking up terrified in the middle of the night. Heather discovered that the baby-sitter had been parking the little girl in front of the TV and showing her a collection of tapes that included *Rosemary's Baby* and movies made from various Stephen King novels. Heather quickly found a new caregiver.

"I've heard Tiffany call her sitter 'Mom,'" Heather says with a sad shrug. "But at the same time it eased my guilt about working—I know she really liked her new sitter."

Heather and Dennis Dixon live in Kingscourt, a working-class suburb that took root east of the huge new Alcan complex during World War II when the company began turning Canadian-smelted aluminum into frames for Spitfires and Lancasters. Dennis himself grew up as a military brat, living in Germany and all over Canada. He was trained as a communications specialist in radio relay at Canadian Forces Base Kingston, moving on to postings in Cyprus and the Golan Heights before giving it up for civilian life. Now the security of a sergeant's pay is gone, along with the military haircut, which has been replaced by a braid that extends nearly to his waist.

For a while Dennis worked on maintenance at the St. Lawrence College of Applied Arts and Technology, where Modern Building Cleaning had the contract. He worked straight midnights, eleven to seven, for "minimum wage plus a quarter," and eventually found himself in charge of a crew of twenty people. But he wasn't happy with the job. Even the low pay wasn't secure. "If you got sick you lost your day's pay." You either worked or you didn't get paid. His crew did all the vacuuming, mopping, and sweeping of a very large building. There was little time for breaks if the job was to be done well.

The Dixons bought their small frame house in 1987, not long after the Frontenac County Board of Education hired Dennis for a $15-an-hour unionized job as a maintenance worker. He has a benefit package "that's a hundred per cent better—because we had no benefits at the College." When he got his first paycheque from the Frontenac Board, Dennis thought there had been a mix-up in the accounting department. "I figured it was for a three-week period rather than the two weeks," he recalls. "I'd paid more out in taxes than what I'd taken home doing the same job out at the college. I thought that when the next pay arrived they'd come back and say 'You owe us $600.'"

Dennis works at the Board's oldest high school, Kingston Collegiate and Vocational Institute, the secondary school of choice for the upper middle class of the city. The building's red-brick construction is the only thing that sets it off from the limestone of Queen's University, which dominates the neighbourhood. He is now the lead maintenance hand at the collegiate. Besides being responsible for making sure the school is kept clean, Dennis acts as security co-ordinator when community groups use the auditorium. He acquaints the users with the building, seeing that they have everything they need for their activities or productions. When he works nights he's the last person out of the building, so he is responsible for checking the alarms; there is no security guard to watch over the computers and other equipment.

Two kilometres down Union Street from KC and Queen's, across the street from the psychiatric hospital, is another of the institutions that help sustain Kingston's economy. St. Lawrence College, built during the boom in higher education in the late 1960s, caters to a student body different from the one at Queen's. St. Lawrence students are generally drawn from the local working class, whereas Queen's attracts far more affluent clientele from across the province and the country. Both institutions, however, use the megamultinational company Marriott, which holds the food-services contract at Queen's and the caretaking contract at St. Lawrence.*

Heather Dixon grew up in Kingston's gritty north end and worked as a waitress before she married. After that she tried a couple of janitorial jobs, which didn't last long: "I won't clean toilets." She has, however, lasted at the Biway. She describes herself as "a nosy cashier." She takes an interest in the customers. When students wearing the Queen's leather jackets come in, she sometimes advises them not to buy two packages of three pairs of socks each but to take the same ones on a five-pair special. She says some of them appreciate the advice but many others simply shrug and say it doesn't matter.

Heather usually manages to average thirty-three hours per week at the store, where she is the only remaining female worker not classified as part-time. Although she earns more than minimum wage, she's close enough to that floor that when the government raised the minimum-wage level slightly, her own hourly rate had to go up as well. All but one of the store's part-timers (the male managers are

---

* In 1994 kitchen staff at Queen's staged a bitter seven-month strike over the low wages paid to part-timers. Queen's administrators claimed that the dispute had nothing directly to do with them.

full-time) are still living at home with their parents. "The girls are all crying for more hours," she says.

She was featured in the Dylex Corporation's quarterly *Biway Bulletin* as an exemplary cashier—although it was only after her smiling photograph appeared in *The Whig-Standard,* the local daily, over a letter to the editor she had written. The headline was "The spirit of Christmas every day." A *Whig* reader wrote in a few days later, "This gal deserves credit for doing her job well. . . . She tries to be the type of cashier she'd like to deal with herself."

Ironically, this public praise appeared only two weeks after a store manager told Heather she had "an attitude problem" and didn't "fit Biway standards." When she got home that day and told the story, Dennis urged her to quit. They could get by on his wages, he said, though Heather insists she can't stand not having money of her own.

"It was good enough for my mother . . ." Dennis protests as they sit on their back porch, which has been turned into a living room. "Your mother controlled the money," Heather comes back before he can finish. Her own mother took in sewing while her father, a mechanic, worked outside the home. As they talk, the couple share puffs from cigarettes that Dennis rolls. The conversation moves on from ground that has obviously been well-tilled to the state of the local job market. An old army buddy of Dennis's who was the best man at their wedding has been unemployed for over two years. He had been working in a shipping room and, like Dennis, had secured a semi-supervisory job. But the job soon disappeared; like many companies, his employers were stripping all levels of management to the minimum, a process known to readers of management journals as "delayering." Once delayered, the friend found nowhere to go in Kingston.

"His wife is part-time at Burger King," Dennis says. "They've given up everything, lost their home. Luckily, they have a very close relationship and seem to be coping." But since the fellow is his best friend, he knows how easy it is for the mask to drop. Such stories are common in town.

"I don't see our economy coming out of this. There's a lot of people like me and my wife just starting—a little late mind you, thirty-two and just starting. If we lose our home, what's going to happen?"

Dennis is classified as management even though he is part of the Canadian Union of Public Employees bargaining unit at the Frontenac Board. I first met him when I spotted a knot of protestors haranguing their provincial representative outside his downtown constituency office. It was only when the NDP government tore up union contracts to impose its "social contract" on public sector workers that Dennis attended his first union meeting. In the aftermath of the dispute Dennis learned that he would have to give up twelve days a year because of his supervisory role; the workers underneath him would give up six, which left Dennis earning about $700 less than them. Now he worries aloud about the possibility that the board might decide to save money by contracting-out its maintenance work to a company that pays minimum wage. "They've got us," he says. Dennis often uses the phrase "putting food on the table."

Heather has noticed that Biway—a discount chain that focuses on the low end of the retail clothing trade—is selling a lot more food than it did when she started working there. Poor people are stocking up on the cheaper brands of canned meat. She has noticed that life outside the store window is getting tougher—one of Kingston's handful of street people was recently killed by a car, she mentions. In 1995 the Dylex Corporation, Canada's largest clothing retailer and owner of Biway, announced that it would close two hundred stores and slash eighteen hundred jobs. Heather Dixon began to ponder the possibility of life outside the store.

In 1954 N.F. Morrison, a Windsor historian, wrote a book about the city at the western end of the lower lakes in which he called the town "the oldest settlement in the province."[1] Putting aside the question of the people who lived there long before René-Robert Cavelier de La Salle sailed into the estuary of the Detroit River, it seems clear that Europeans had already arrived at the eastern end of the Lower Great Lakes before they made it to Detroit. In fact, immediately after Count Frontenac first met with the Iroquois at Fort Frontenac in 1673, he granted La Salle seigneurial rights and a trading monopoly in the area. It was its military and mercantile value as a fur-trade outpost that gave the settlement its initial value and La Salle, a shady intriguer who had no long-term interest in the place, would use Kingston as his base for further explorations. Due to its hardscrabble hinterland and the attitude of the Six Nations people, it never really took off as an agriculturally based settlement.

The British Loyalists at the mouth of the Cataraqui were as worried about the Americans as the French had been about the Iroquois. They changed the name to Kingston and, recognizing the strategic nature of the juncture of Lake Ontario and the St. Lawrence River, built another fort after the War of 1812. They called this one Fort Henry. Its sturdy stone parapets were never used in defence against the Americans and instead sheltered several generations of nineteenth-century remittance men, officers in Her Majesty's Army. Fort Henry now overlooks the Royal Military College, the training school for Canada's officer class. Surrounded by the Canadian Forces Base (CFB) Kingston, one of the biggest military bases in Canada, Fort Henry has long been one of Kingston's prime tourist attractions. True to the age, it now advertises itself not as an old fort but as "Ontario's First Theme Park."

Following the war with the Americans Kingston developed as a centre of trade and administration. A kind of entrepôt trade that its prosperous merchant class called "forwarding" sprang up, with lake schooners offloading the surplus of the Upper Canadian countryside (timber, grain) onto batteaux and Durham boats for the journey downriver. Kingston shared with many ports a reputation as a rough town; sailors engaged in regular punch-ups with the soldiers of the garrison in the taverns along the shore. What it did not share with most other ports were the imposing public buildings that came into view as the thirsty sailors approached Kingston.

In the early 1840s the town served briefly as the capital of the

United Province of Canada. Its grand City Hall was built to serve as the seat of government. Along the shore at Portsmouth loomed Upper Canada's first penitentiary, described by Charles Dickens as "an admirable jail . . . excellently regulated in every respect."[2] The Kingston Penitentiary was soon joined by another foreboding institution, the provincial asylum. In 1841 Queen's College was established by the Presbyterian church to train young men in the ways of the ministry. All of this early building put its stamp on Kingston. Although Lord Durham called it "a highly important station, both in a commercial and a military point of view" in his famous Report, Kingston became known as an institutional town.[3]

Perhaps no one has understood the story of Kingston better than Arthur Lower, the historian who, with Harold Innis, elaborated the staples theory to explain the country's development. Lower lived in Kingston for forty years and described it as a "sub-capital," a place where the functions of the state are crucial. The city, always a major military centre, received a great boost with World War II when two dominant multinational firms, Du Pont and Alcan, established nylon and aluminum plants there. Rapid population growth followed. "All of Canada has benefitted (economically) from every war, first and last, in which it has been caught up," Lower observed. "And no town has benefitted more directly, positively, proudly, and righteously, than has Kingston."[4]

After the war came an accelerating expansion in the size of government and its workforce. Nylon for parachutes became nylon for stockings, and aluminum foil and siding found ready consumer markets. Military budgets remained solid. And the growth of education and the welfare state meant growth for Kingston, with its three large hospitals, its university, and its military college. The prisons kept pace. With its eight local jails, the city is known as the "prison capital of Canada." Visitors often wonder about the function of a turreted, scarlet-roofed structure that sits on ample grounds close to a main thoroughfare and looks—at least from the outside—like a Disneyland version of every kid's fantasy castle. It is the Collins Bay pen. Newer, more up-to-date jails like Millhaven, a maximum-security facility nicknamed the "Gladiator School" by its staff, have been tucked discreetly out of the view of tourists passing along The Loyalist Parkway west of town. All of this means that Kingston has enjoyed a measure of affluence, particularly after the wave of public-service unionization that began in the 1960s. Jail guards, hospital orderlies and nurses, office workers at Queen's and on "the base," clerks at the new headquarters of the Ontario Health Insurance Plan: they all saw their wages increase. The importance of subsequent unionization was underlined in 1965, when the province announced an increase in the general welfare rate; Kingston hospital administrators expressed immediate "dismay," claiming that if the welfare boost went through, some low-paid hospital workers could find themselves earning $600 per year less than welfare recipients.[5]

A welfare-dependent population has always been notable in Kingston, a town fractured along class lines. Until recently, Princess Street was a very real dividing line between the comfortable and the poor. No merchant, physician, retired officer, or even a member of the burgeoning administrative middle class would think of living "north of Princess" where, as late as the 1960s, there was still housing without sewer and water service. Lower attributed the "segregation" he described in Kingston's north end to the drift of newcomers from the rocky farm plots of eastern Ontario: "Poor land makes poor people."[6] Add to this the presence of prisons that bring poor people to town as inmates, with their families following, as well as the early influx of famine Irish in the 1840s (twelve hundred of them are said to have been buried in a mass grave near Kingston General Hospital) and you get an idea of why Dickens's description of Kingston as "a very poor town" still has resonance. As late as 1970 the rate of working-class home ownership in Kingston was 47 per cent, compared to a rate of 68 per cent in the country as a whole around the same time.[7]

This is not to say that Kingston is the same today as it was in the 1950s when then-resident Robertson Davies wrote a satirical trilogy about an all-too-familiar town he called "Salterton" where dowager ladies rattled around Victorian mansions sipping sherry and eating prunes wrapped with bacon. Those mansions are still to be seen, "old stones" that attract both tourists and new buyers who have moved into old houses and buildings—even in the north end—to fix them up in a process of gentrification. The economy and with it the labour market have changed as well. Old industries that bespoke the city's former role as a transportation hub and dominated the waterfront are gone, replaced by yacht clubs and condominiums. The works that produced the largest locomotives in the Empire and the first diesel-electric locomotives in North America shut down in 1969, and the shipyard closed in 1968.

Doug Tousignant, who has been everything from die corrector at Alcan to cafeteria helper, recalls Kingstonians regularly setting their watches by the noon whistle from the locomotive works. His family history and forty-year career symbolize the trajectory of Kingston's working class. His father was an engineer on the lake ships, and two of his seven brothers started as machinists at the locomotive works. When that closed down, he says, men could still take their skills to Alcan, the biggest employer in town in the immediate postwar period. Doug's wife Bonnie worked doing invisible mending at the old riverside knitting mill where Dominion Textile and Hield Brothers produced worsted fabric for nearly a century before the place was closed. Later the site reopened to house an upmarket restaurant, architects' offices, and a new-age health centre.

Doug can rhyme off the names of all the other small manufacturers that supplemented the jobs at the big Du Pont and Alcan plants: the tannery, the tile works, the picture-frame plant, Kingston Spinners, a half-dozen others. Although Kingston was never a dynamic centre of manufacturing, its partially diversified industrial base must have been the envy of many centres in northern and Atlantic Canada, where in the 1990s fifty thousand fishery workers have joined the army of unemployed people trying to cope with fear and feelings of uselessness while being advised to run whale-watching tours or dispose of garbage from the United States.

"The Williamson plant is the only one still operating at Alcan, and it employs about four hundred people," Doug says. When he started in 1948 Alcan's extrusion department *alone* had 427 workers.

A look at local employment trends between 1970 and 1989 gives a sense of what the retired metalworker means. In that period the employment level at Du Pont, which was by that time the biggest industry, dropped by 48.7 per cent. Employment at Alcan, the number two industry, declined by 56 per cent. The largest overall employer, CFB Kingston, registered a 36 per cent increase. Queen's University displaced Du Pont as the second most important employer in total, boosting its workforce by 60 per cent. Corrections Canada (the polite name for the jail business) also saw its payroll expand by 60 per cent, which brought it up into the number three position.[8] By 1995 Queen's had surpassed the Base to become the largest employer in town.

These figures may not reveal broader trends like the growth of part-time work in both the private and public sectors. They also include all employees—management, labour, and professional—and ignore the arrival of new employers like Northern Telecom's cable plant and the OHIP headquarters, each of which initially created about six hundred jobs before the apparently inevitable "downsizing." But the trend is still clear. The service sector, particularly that represented by government employment, has become overwhelmingly dominant. Indeed, like elsewhere, manufacturing jobs dropped sharply in the recession of the early 1980s. By 1989 they had crawled back to pre-recession levels, only to plummet again.

All of this is obvious to the Kingston Area Economic Development Commission, which sees a future tied firmly to the service sector. Doug Tousignant, who spent his paid working life dealing with tangible things like aluminum ingots and dies and now, in retirement, concentrates on a meticulously cultivated garden, has trouble with this.

"They can't *all* be service centres," he says—meaning the towns dotting the north shore of Lake Ontario, all of them seemingly with plans to cash in on their historic appeal. "This week we had a group of people giving away pancakes down at the waterfront and telling us tourism is our biggest industry. I hope to God it isn't. I'm very proud to be from here. It's without a doubt one of the most beautiful cities in Canada. I know we have a lot to offer. But it's gotta have more than some parks and a fort and an ancient city hall. How often can you visit a fort?"

Like so many other deindustrializing centres—from Glasgow and its museums to Windsor and its casino, Kingston is staking a good part of its economic future on its potential to attract visitors and their money. Each spring the Economic Development Commission and the Chamber of Commerce kick off the tourist season with an outdoor breakfast in a small park between the former legislature of the united Canadas and the Flora MacDonald Yacht Basin. Rotarians serve up pancakes beside the old railroad station that now serves as the main tourist information office. A huge black steam locomotive donated by the Jaycees and named "The Spirit of Sir John A." after Kingston's most famous son stands nearby. The leather-lunged Official Town Cryer in full regalia shouts out a few "Hear Yeas!" while women in period costume hand out pamphlets advertising the Bellevue House National Historical Site (a house John A. Macdonald occupied briefly), the woodworking museum, the prison museum, the maritime museum, the military museum, and the international hockey museum. There is a clear emphasis on the old-fashioned.

Meanwhile a handful of civic worthies in smart suits mingle on the lawn. One talks into a cellular phone. The local Easy Favourites radio station is on the scene, "doing a remote," and it uses the event as the lead on its eight o'clock news: "Kingston's biggest industry is being spotlighted as Tourism Awareness Week kicks off with a pancake breakfast . . ."

Kingston planners are proud that tourism has become a "managed industry" thriving on the co-operation of the private and public sectors. Tourism Awareness people say the business is responsible for eighty-five-hundred jobs in the area. Paula Nichols, who has a full-time job co-ordinating tourism for the Development Commission, admits that although there is more employment in the institutional sector as a whole, the tourist business is a bigger job source than Kingston's fading manufacturing base.

"We promote history and water-based recreation," she says, describing the management strategy. "The fact that it was Canada's first capital; Sir John A. Macdonald is buried here; the limestone architecture. The water-based recreation is the Thousand Island cruises, the sailing, and the waterfront. It's the quiet old place with modern facilities. Shopping is also a key attraction."

I tell her about Doug Tousignant's critique of tourism, that it creates low-wage jobs that are often temporary, part-time or both. "The starting positions with any service business are lower in salary," Paula Nichols explains. "There's a need for that kind of employment. And not all of tourism employment is like that. There are a lot of managerial and marketing positions that people work their way up to. . . . If we want to throw away the eighty-five-hundred jobs, then we can. But it provides the diversity that a healthy economy requires."

Kingston is assuredly an old town: the word "heritage" is repeated four times in the first two paragraphs of its main tourist brochure. But oddly enough, the emphasis on tourism is recent. As late as the 1970s the downtown area had not yet been gentrified and was still down at the heels and disreputable. Helen Cooper, a former mayor who went on to head the powerful Ontario Municipal Board, recalls girlhood walks downtown to meet her father, who worked at the Whig-Standard Building behind City Hall. For her, the last few blocks were unpleasant at best: "I hated every minute of that. I didn't feel like I was going to be attacked, but the atmosphere was so seedy and unpleasant—grungy stores and a sleazy atmosphere." On one corner where sidewalk cafes now thrive in the summer tourist season, the ex-mayor says there was slum housing where twenty families lived. "It was like something out of Victorian Britain."

As the downtown was deindustrializing, people from the professional middle class, mostly liberals associated with Queen's University, bought up handsome properties in the area between the university and Princess Street and began to pressure the city to preserve Kingston's historic character and reject various megadevelopments proposed for the waterfront. This articulate, well-organized group had a measure of success. Although its members failed to stop several heavily criticized projects, their agitation was enough to deter some grandiose schemes, and much of the historical character of the townscape was preserved. This helped set the stage for Kingston's re-emergence as a heritage tourist centre and, more importantly for the hopes of its planners, the growth of what the Economic Council of Canada calls "dynamic services."

On the local level this type of growth feeds into dominant thinking about the future of the economy and, by implication, the labour market. Dave Cash, executive director of the Development Commission, is convinced that Kingston is "out in front," that it is shifting quickly in the direction of "where the economy is going."

In 1992 the city received a report it had commissioned on the future of its economy. According to the report, Kingston's main strengths included its "quality of life" and "local heritage." These, together with the existing "human resources," meant that Kingston was well-positioned to take advantage of an emerging high-tech, knowledge-based economy. This analysis accepted the conventional wisdom of the day: a city's success would be determined by "its perceived total performance in responding to business needs"; the need for life-long education because of high unemployment among the less educated; the need to get away from "big government and big industrial establishments" and move to "a future based on traded goods and services, characterized by more and smaller entrepreneurial [sic] driven firms." The report identified computer software, research and development, biotechnology, and group tourism as some of the most promising target sectors and highlighted Queen's as "one of Canada's most valued sources of professional, entrepreneurial, and managerial talent."

"The hope for Kingston, and for Ontario as well," concluded consultant Stephen Chait, "is not to pursue a low-price, assembly-related, limited service industrial and service role, but rather to shift resources and talents toward more modern, higher value-added per employee activities and facilities which also pay higher wages and increase the standard of living of the community as a whole."9

Using less tortured phraseology, Queen's geographer Maurice

Yeates summed it all up in a lecture entitled "Kingston and the Future: A 'Clever' Community?" People, particularly the clever folk whose work is a sparkplug in an era when wealth is created by knowledge, prefer living in pleasant environments. Kingston, despite the fact that its waterfront was "re-appropriated by private interests" after the end of the commodity and manufacturing eras, is "better placed than many other Canadian communities with respect to the new global economy." Why? Its educational institutions and ultra-modern hospitals offer a research and education base well suited to a new postindustrial era.[10]

One of Kingston's biggest economic development announcements came in 1993 concerning the highly touted Queen's University bioscience research centre. The event brought two top provincial cabinet ministers to town. Although the $50 million facility will generate limited direct employment, it is exactly the sort of thing that many see as the wave of the future. The centre is to be an innovative place that could spawn new products like insulin-producing bacteria or new pesticides, products that can in turn spin off new businesses. During a period when universities are under pressure to make their research applicable to commercial endeavour—and indeed to serve the needs of business more directly than in the past—this is a sign of the times. So too is the fact that the university is looking to private corporations to assist in equipping the new labs.[11] The Queen's bioscience centre will also include a Technology Transfer Centre staffed in part by people from the local Development Commission. It is hoped that this offshoot will facilitate the emergence of new businesses.

The announcement produced a blizzard of adulation on the part of the local press, which deployed the usual adjectives: high-tech, leading-edge, state-of-the-art, and, of course, *excellence*. A key backer of the whole project was Bob Burnside, the vice-chair (and later chair) of the Kingston Area Economic Development Commission as well as a former Imperial Oil executive and one of the top bureaucrats at Queen's.

"Kingston is already an important technology centre," said Burnside, who also helped to get the Greater Kingston Technology Council under way. "The technology Council will help us to develop spin-off businesses, attract new high-tech industries and create high-paying jobs." Frances Lankin, the NDP minister for economic development, added, "This does create jobs—now and in the long term."[12]

Jack Gilbert (not his real name) has worked for several years at a firm that is the perfect model of success in what Kingston wants to be. Before he was hired at QL Systems Ltd., Gilbert—twenty-seven with an undergraduate degree—could only find work at a telemarketing outfit.

"Here I was, this apple-pie-faced, conservative-looking person, and I couldn't get a job. Imagine what it's like for someone who doesn't fit the mould. I ended up telemarketing tickets to variety shows. It was quite disturbing. I began to have more empathy for the people robbing variety stores."

Gilbert was virtually computer illiterate when he started at QL, and it took him two weeks to learn the job, mostly by himself, mostly by trial and error. He had heard about the job from a friend, whose boss had asked, "Can he read? Can he write? Send him in." Gilbert works the four p.m. to midnight shift, and he can wear what he likes to work. On the job he uses a scanner to transfer typed copy into a computer, and then he compares the original to the inputted material (scanners are one of those time-saving computer devices that are as yet far from perfect). He cleans up the spelling, syntax, and typos using his own knowledge of grammar and the program's handy spell-check function. His job is formally called "editor." Gilbert says, "It's a nice title but it's just copy-editing."

Half of the sixty people employed at QL do similar tasks. During the half-hour dinner break, Gilbert can sit on QL's waterfront property and take in the view, with sailboats in the foreground and the green copper roofs of the military college beyond. Sometimes hungry but confused tourists wander up to ask if the board and batten building that houses QL is a restaurant.

QL is owned and run by Queen's law professor Hugh Lawford, who came to town when the university's law school opened in the 1950s. The firm emerged from a joint IBM-Queen's project on information retrieval and now specializes in servicing lawyers across the country by providing instant, on-line access to judicial decisions. For this work, Lawford insists, speed is crucial. "The Supreme Court releases its judgments every Thursday, and we have them available across Canada within an hour of their release by the court."

Since its 1973 incorporation, QL has expanded to provide scores of databases dealing with everything from law and business information to hydrology and mining technology. It also markets software packages and provides contract programming and consulting services. Lawford's firm is proud of its global network of users in at least twenty countries and its string of offices in Toronto, Vancouver, Calgary, and Halifax. "The databases operated by QL," states a company handout, "with their associated indices, currently contain more than 48 billion characters of information and are being expanded."[13]

QL systems is without doubt technically impressive, with its mainframe computers and access to nine hundred databases. It needs two kinds of workers: people who can trouble-shoot computer problems and do programming and others (like Jack Gilbert) who can come in and learn the job quickly with a minimum of training. According to Lawford, his firm is "increasingly moving in the direction of hiring only university graduates because they come in and are very flexible and can move into a whole variety of tasks." Advertising recently for an editor, the company got 185 applicants. Lawford, clearly proud of the fact that half of his top managers are women, points out that there are few spatial limitations in the business. The editor-in-chief moved from the Vancouver office to the Calgary office before she settled in Halifax. He is also pleased with the way his business fits into Kingston.

"Here we have a service industry and no one even notices us," he says, wheeling his leather chair from desk to computer in the oak-trimmed executive office. "We don't pollute the environment. . . . Where else in the world could I buy a building for my office that is surrounded on three sides by water and has a view like the one you can see here? I can walk to the office—it's within a half mile of the university."

The building in which QL Systems reprocesses information before selling it to the on-line market used to house a marine towing and salvage operation in the days when Kingston had a working port. It is located close to the spot where Doug Tousignant's father used to board the pilot boat that would take him to work in the engine room of a laker steamer. A block and tackle hangs as decoration from the vaulted ceiling of Lawford's office. An old Bakelite rotary dial phone sits on a reception counter which, like the rest of the building, is finished in rich pine and oak. The place oozes with nostalgia for a past era, although Lawford's right-hand woman, Kingston native Lillian Simpkins, describes the old, long-gone waterfront as a "pretty gruesome" place. Her dad used to work at the locomotive works to supplement his income from farming. Nonetheless, QL has attempted to re-create at least the flavour of that past: the work space is lit by custom-made imitations of the one remaining porcelain fixture found on the site before renovations began. "We wanted to maintain the spirit of the old building," Lawford says.

The view from the word processor is, of course, different. The work Jack Gilbert does is not "high-tech," he says, though the "mainframe work" might be. Gilbert describes QL Systems as "a word factory" where adherence to industrial discipline is "quite rigid." In spite of the casual clothes and elegant surroundings, he thinks the place should have a punch clock. The "peons," as he calls them, are not given the whole picture of how the business works. Employees have to pledge not to reveal their rate of pay to each other—you can be fired for telling, Gilbert says. "I don't know why. But you immediately think bad things about it and assume they're pulling a fast one on someone. It ensures you can't get a union and that you can be screwed and not know it."

Despite this drawback Gilbert says that none of his fellow keyboard jockeys would say it is the worst job they ever had, though morale is not particularly high. He has his sights set on graduate work.

What remains in this highly computerized workplace from the days of sail and steam are the social relations. Gilbert is the ideal worker for an apparently new era. He is flexible enough to learn the job by himself, and he is replaceable if need be. He brings few expectations to the job beyond the desire to earn money. Modern managers who emulate Toyota's much-imitated "just-in-time" system of inventory control have also embraced the notion of an ample inventory of just-in-time workers—ideally pre-trained—who can be slotted in on a temporary, contract, or part-time basis.

Have computers brought the kind of change that mesmerized futurists would have us believe? Do new-age companies that prosper by deploying computer technology represent a radical break with the past? Artificial intelligence pioneer Joseph Weizenbaum of MIT says, "Yes, the computer did arrive 'just in time.' But in time for what? In time to save—and save very nearly intact, indeed, to entrench and stabilize—social and political structures that otherwise might have been either radically renovated or allowed to totter under the demands that were sure to be made on them."[14]

Of course, this does not mean that changes are not being wrought to the social structure as a result of new kinds of production of new sorts of commodities like databases and genetically modified fish able to survive in acidified lakes. Asked what kind of society he sees emerging, Hugh Lawford pauses for a moment, then describes a dual-labour market containing "the kind of people who work selling hamburgers at McDonald's on the one hand and the university people on the other, *with not much in between.*"

———

On March 2, 1990, Garrie Manser punched out at the north-end picture-frame plant for the last time. K-D Manufacturing had closed and moved across the river to Watertown, New York. Manser, with twenty-one years as a painter at K-D, remembers the exact date three years after the fact. He can also remember his first day at K-D, November 13, 1968. He was twenty-two at the time.

Manser, raised in foster homes in the Kingston area, is a slim, wiry fellow with a full head of greying, curly hair, long sideburns, and an unassuming, even hesitant manner. Since his painting job was exported he has spent much of his time on unemployment insurance or welfare. When he first went to the local Canada Employment Centre, a counsellor advised him to prepare a resumé. "I didn't bother about that," he said at the time. But his daughter soon prepared one because the crowds applying for the same jobs were flooding prospective employers with paper.

After K-D shut down Manser managed to find work with Dust-busters (a contract cleaning outfit) and the Ministry of Transport. He liked the government job; it involved painting highway guardrails, and he could use his experience. But both jobs ended soon after they started. He also got in six weeks at a door and window maker, but left because chemicals in the wood preservative gave him an unbearable rash. Construction is out of the question because of back problems resulting from lifting at K-D. His wife Marjorie quit her kitchen job at Kingston General Hospital after they were married and has not held regular paid work since. She brings in some money doing home child care, but Garrie Manser seems convinced that she will "never get a job." Their daughter Dianne, who has completed a course at St. Lawrence College, had a contract job doing graphic design work at CFB Kingston, but that ended and she had to move to Cornwall to find work. Their son has had temporary work as a landscape labourer at "the base."*

"There's so darn many people out of work, they don't know who to pick," Garrie Manser says. His twenty-one years working in one place doesn't seem to carry any weight. "They want a Grade 12 education. The jobs you see in the paper are for bartenders, cooks, and like that. They want experience."

---

* In many Canadian communities with military bases—and even more so in the United States—the armed forces are a fact of economic life for the working poor. The military has traditionally offered at least temporary employment, which accounts for the political sensitivity of base closures.

Although Kingston has more than its share of labour-market experts, what with the business and industrial relations programs at Queen's, the town's most street-wise, savvy analyst lives in a north-end apartment crammed with files and copies of the most recent changes to the regulations governing the unemployment insurance system. Before she became a full-time—though often unpaid—advocate for the unemployed, Theresa Houston worked in nursing homes and made sandwiches for a mobile caterer that served construction sites and small factories.

Houston has what might be called a "vertical" perspective on the labour market. It is a society-as-hourglass view, and she sees many people being filtered downwards. The world according to Houston has its "niches"—people sometimes fit into them, but they can also be dislodged from them. She bases her vertical perspective on twenty years of dealing with people looking for work in Kingston.

"The people who used to clean theatres, sling beer, drive cabs—they were the very poorly educated people," she says in a voice that retains more than a little of its Scots brogue. "They had a niche." Now what's happening, she says, is, "The people up above, the better educated and the Queen's people, started filling in those jobs. Nearly every barman is from Queen's. So we have this great displacement." Unlike the original "DPs" (Displaced Persons) who arrived in Canada from the smouldering ruins of postwar Europe, many Kingston DPs of the 1990s have already migrated to service and small factory jobs.

Having been displaced from K-D Manufacturing, Garrie Manser has had trouble finding a new niche beyond a series of on-and-off jobs and welfare. According to the conventional wisdom, he is not flexible enough for a modern labour market that demands adaptability, the learning of new skills, and the ability to roll with the punches. Then, too, his social identity and self-esteem are strongly wrapped up in his long years of experience as a painter. Like many people, he identifies with his working skill. He takes pride in it; he is reluctant to turn away from what he knows best. I once asked him about the training programs available for the unemployed.

"There was nothing I was really interested in," Manser said.

There are, of course, those who would say that Manser is not living in the "real world," that the age of falling expectations is all about competition: winners and losers, the skilled and the unskilled. One of those would be University of Chicago economist and Nobel laureate Merton Miller, who was asked by *The New York Times* in the days before the final congressional vote on NAFTA to address the concerns of U.S. workers worried about their jobs migrating south—just as Garrie Manser's had. What about all the steel and packinghouse jobs that had brought a measure of prosperity to Chicago workers? Miller replied, "It's not Mexico. It's the machine." Free trade or not, the argument goes, corporations will relocate and invest in new technology. "A lot of people are worried about keeping high wages for low-skilled jobs," Miller insisted. "And they should worry. In the modern world, those two things don't go together."[15]

It is not that Garrie Manser was earning high wages at K-D— $11.95 an hour provides a *gross* income of under $25,000. When he leaves his one-page resumé with possible employers, they see that he may have just barely made it into a category that, according to UNESCO, indicates functional literacy for people in industrialized countries. Less than Grade 9 is defined as functional illiteracy, and Garrie Manser's resumé says he completed a two-year "Occupation Course" in high school. Almost 3.5 million Canadians (or 17.3 per cent of the people over fifteen) have not reached that stage.[16]

What does the future hold in a competitive, high-skill job market that, we are told, is here to stay? What are the possibilities of latching onto one of the vast number of low-wage jobs at the bottom of the hourglass?

The welfare office requires a job-search in exchange for a welfare cheque: two contacts a day for Garrie Manser. "They wanted me to get the paper full so they held my cheque back," he says. "I called welfare and the caseworker said I didn't have the job search full. I said it was three-quarters full. 'Well, we want it full,' she told me." The Mansers live twelve miles out of town and Garrie spends his days dropping off resumés and checking out leads. "Can you imagine the amount of running around I gotta do? My truck is small but it still takes gas."

In the end Garrie Manser was finally forced onto the training bandwagon. He landed a job—subsidized by public training dollars—painting and staining furniture with a small local business, a tiny workplace of four people. This new shop essentially does the same kind of work he performed at K-D but uses a greater variety of colours and tints, so Manser is glad to have the initial period of training. His pay is 16 per cent lower than it was at K-D.

"It just doesn't make common sense any more to job search," admits Bernie Mason, the chief administrator of social services for the city of Kingston. Between 1989 and 1993, he says, there was a net loss of between twenty-five hundred and three thousand jobs. On a good day in 1993 there were about thirty jobs going. If just 10 per cent of the caseload were actually searching for work, they would drive employers crazy. Although he shares the prevailing wisdom that the future lies in tourism and high technology, Mason is more sanguine than most senior officials when it comes to the realities of the new world of work. "People don't think what we're in is a Depression because we have the welfare system," he says. "And welfare is becoming the new UI system."

Kingston has traditionally carried a welfare caseload of between nine hundred and twelve hundred, with most of the recipients geographically concentrated north of Princess Street. It was all very clear, like something designed by a sociologist to prove a theory. That was the area where the poor of Kingston had always lived, trapped in a generational cycle that bred despair and violence. The situation was not helped when, in the 1960s, residents of more affluent parts of town resisted the location of public housing in their neighbourhoods, with the result that the poor were further ghettoized in the north end, where it was built.

In August 1989—some seven months before K-D Manufacturing closed—the pattern that had held since the 1960s began to shift. When Mason saw his caseload hit twenty-eight hundred he began talking about a "*nouveau* poor" emerging. He attributes it to the shift in the job market as more people lost what had once been seen as good, secure jobs. The people on welfare are now spread all over town. By late 1993 Mason was budgeting for a caseload increase of between 5 and 8 per cent." Because of the sheer numbers the traditional geographical pattern of general assistance cases has been destroyed," he says. "Now it's like tossing a handful of corn onto the map of Kingston."

One of the newest, best-appointed buildings at Queen's University is the School for Policy Studies.

This only makes sense. Queen's is an elite institution that has employed and trained some of Canada's most powerful men, including many of the civil servants who have shaped public policy. According to George Grant, "The officials of the Department of Finance had mostly learnt their economics at Queen's University in Ontario, where the glories of the free market were the first dogma."[17] Economist Clifford Clark, who laid the groundwork for the Bank of Canada, became the first head of Commerce at Queen's in 1919 and taught special courses that the university offered to bankers. He later became deputy minister of finance. A personification of the close and natural ties between business and government, Clark was employed for a time by a Chicago firm that financed real estate speculation. He had a particular interest in unemployment and, with his Queen's colleague Bryce Stewart, set up the system that tracks unemployment rates. One of Clark's mentors was another Queen's man, O.D. Skelton, usually described as the founder of Canada's modern civil service. A Presbyterian and "staunch believer in liberalism and individualism," Skelton began his career in 1911 by writing a book attacking socialism; he ended it when he died on the job as undersecretary of state for external affairs.[18]

In 1993 Queen's invited Professor Lee Dyer, Director of Cornell University's Centre for Advanced Human Resource Studies, to its Policy Studies complex to deliver the annual Don Woods Lecture in

Industrial Relations. Dyer, a leading U.S. theorist in the field of personnel management, began his talk to a packed hall with an anecdote recounting an incident from one of the many seminars he conducts for managers at organizations such as IBM, Mobil Oil, the U.S. Navy, and the National Research Council of Canada. One harried boss looked at Dyer and, in what the professor described as a desperate-sounding voice, asked, "When are things going to return to normal?"

"The mess that we're in now," Dyer told the man, "will be with us for some time."

The "mess" under discussion referred to changes in how large organizations do business, the moves towards "flexibility" in both mass production and organization. It all has to do with managers (and often workers) working in teams, the supposed elimination of hierarchy within the firm, and the need to be able to respond quickly to a constantly changing business environment. Such an approach to shaking up the world of business is often advocated by McKinsey & Company, one of the world's most influential management consultants. Tom Peters, a McKinsey alumnus, has packaged his experiences into a hugely successful career as author and management guru. The flavour of the times is illustrated by the very title of Peters's 1992 book, *Liberation Management: Necessary Disorganization for the Nanosecond Nineties.* The emergence of both management consulting and, more recently, management gurus may have something to do with the explosion of credentialling programs like MBAs and the fact that managers have been busy developing a strong sense of self-awareness—or self-absorption. It may be that as the world of business changes as fast as the rest of the world, people in power seek reassurance. Whatever the case, corporations big and small are willing to pay huge fees to a new coterie of operating prophets to advise them. The huge success of people like Peters, Michael Porter, Peter Sedge, and (there is even a woman) Rosabeth Moss Kanter is proof positive that the service economy is not just about low-end jobs as burger-flippers.

In this context Lee Dyer's message to Queen's was scarcely surprising. He spoke of a new "paradigm" and the end for the old tall, pyramid-shaped corporate organization, which must give way to "adhocracies . . . flat, more organic, flexible organizations" characterized by "flexibility, speed and innovation."

"Everything that an organization can't do by itself better than other organizations should be *outsourced*," Dyer advised. This is precisely what is happening across the economy as private businesses and government departments contract-out their operations to smaller (often non-union) firms that employ people on a contingent, just-in-time basis.

What was perhaps most interesting about Dyer's lecture was what he did *not* focus upon, particularly since Industrial Relations (or "IR" to the cognoscenti) has historically been about how managers deal with their mass of workers. For Dyer (and many other corporate planners) the crucial people in the new lean organizations are not the easily replaceable folk who can be plucked from the swelling ranks of the unemployed. Rather, profits will depend on those he calls the "P&TS"—professional and technical employees. Blue-chip outfits that consult advisers like Dyer are less interested than they once were in how to deal with their blue-collar workers. They regard one of their main challenges as attracting and keeping the skilled designers, engineers, and systems people who are seen as the key to competitive success in the information age.

Indeed, Dyer's lecture made only passing reference to what he stereotyped as the "downtrodden proletariat." Such people are the ones whose labour has been "outsourced" while "those people *who continue to be employed*" are "upskilled." Dyer was almost wholly concerned with these P&TS. It seems that, contrary to the Canadian Manufacturers' Association economist who said that there is no job that is secure any more, there *are* indeed jobs these days that offer good pay and reliable employment.

If you want one of those good jobs, according to the powers-that-be, it is crucial to become "upskilled" and to enter the ranks of what Robert Reich calls "symbolic analysts." These are service-sector people who work at the top of Theresa Houston's vertical hourglass. What do they do, these symbolic analysts, the people said to represent the wave of future—individual—prosperity? Theirs is a world of symbols, representations of data and words. According to Reich:

> Included in this category are the problem-solving, -identifying, and brokering of many people who call themselves research scientists, design engineers, software engineers, civil engineers, biotechnology engineers, sound engineers, public relations executives, investment bankers, lawyers, real estate developers, and even a few creative accountants. Also included is much of the work done by management consultants, financial consultants, tax consultants, energy consultants, agricultural consultants, armaments consultants, architectural consultants, management information specialists, organization development specialists, strategic planners, corporate headhunters, and systems analysts. Also: advertising executives and marketing strategists, art directors, architects, cinematographers, film editors, production designers, publishers, writers and editors, journalists, musicians, television and film producers, and even university professors.[19]

It could be that not everyone in this dog's breakfast of the professional middle class enjoys job security. Many (including this writer) work for themselves and are subject to the ebb and flow of the market for their services. Dyer predicts that two out of every five of his P&Ts will be working on a temporary, part-time, or contract basis. If you can get this kind of work, if your skills are in demand, it does offer a certain autonomy; perhaps this accounts for the popular appeal of starting a small business and why so many people attempt to do so against all odds. But the point is that the private bureaucracies dominating the economy and the public bureaucracies performing a myriad of regulatory and service functions need the labour of many such people. They are prepared to pay well for much of it. According to Dyer, "the bottom line" is the "need to offer employment stability." This is particularly true of jobs that require a lot of "discretionary effort" (which goes to say, a high level of commitment coupled with the willingness to work long, long hours).

66

This type of work, of course, offers little or no relief for the people at the bottom of the hourglass. In his Queen's lecture Dyer passed over what he calls "support" people, whose relationship with the firm is "tangential." He did, however, ask himself a rhetorical question about the average pay of all new jobs in the United States. "Six bucks an hour," was his answer. He admitted that it "doesn't meet the poverty line in the U.S." Nor does it in Canada.

There are, of course, adherents to what John Kenneth Galbraith has called "the quiet theology of laissez-faire" who argue that the best protection against poverty is a job—and this is surely the case if one can find a job as one of Reich's design engineers, investment bankers, or marketing strategists. But for more and more people—particularly those who start at the bottom in a place like the north end of Kingston and cannot dream of ever getting into an institution like Queen's—such a view is simply a cruel hoax. It is not merely a matter of living in poverty, it is about *working* in poverty. The Canadian labour market has not produced a decent living standard for everyone. In fact, it is giving rise to an ever more unequal division of wealth.

This is what the age of falling expectations is all about. It plays itself out in a number of ways, but especially in people's hopes for their children, and in children's expectations for themselves. A survey of North Kingston youth taken in a relatively prosperous year (1988) revealed that 22.9 per cent of the sample were either unemployed or had been so in the past year. About 90 per cent of respondents between the ages of twenty and twenty-four were looking for full-time work, and young women were 1.8 times more likely to be working part-time than full-time. When asked what they felt they would be doing in ten or fifteen years, 57 per cent of youth from the public housing concentrated in the community replied either that they had no idea or that they would be doing the same thing.[20]

Five years later, as the stubborn recession hung on tenaciously, *The Whig-Standard* published an editorial cartoon showing a child sitting up in bed in a frightened sweat, his teddy bear beside him. He had been counting sheep. The cartoonist's balloon pictured two dead sheep that had smacked into a large wall. One was labelled "Jobs," the other "Environment." A third sheep labelled "Future" was just obliterating itself against the stone. The caption bore the simple inscription, "Report: Our Children Have a Dream."

The report in question was the result of an Insight Canada Research focus group study commissioned by the Premier's Council of Ontario. The pollsters had set up twenty-two sessions (half with parents, half with children) in what had historically been Canada's most prosperous region. Participants were screened to achieve a sample of class, ethnicity, and demographics. The leaders opened the discussions with a general query about what children "hope and dream about."

The results were startling and depressing, particularly in light of the prevailing emphasis on competition in the skills sweepstakes. They were also sometimes poignant.

"There should be a program in the schools to at least give them something to eat in the morning," said a twenty-one-year-old Salvadorean refugee and high-school dropout. "At least if you've got something to eat, you can learn better; after that, if you want to learn, it's up to you.... Your mind works with your body and if you don't feed your body your mind won't work that well."

The principal conclusion of the study was that kids suffered from a dearth of confidence and hope that their aspirations were achievable. Students were reported to be "acutely aware" of the climate of insecurity in which their parents lived. "In today's Ontario, the self-perceptions of working class and middle class families have merged and taken a 'downward' turn: people fear for their jobs, live from paycheque to paycheque, save little, are often a paycheque away from familial insolvency, have little hope for change and find it difficult to conceive of, let alone place their hands upon, the levers that will help them invoke positive change in their lives. Children live with this."

This was the dominant "tier" the investigators noticed in a framework of top-of-mind perception. The second tier related to broader, more philosophical aspirations, including freedom and self-sufficiency: "Many young people crave being able to support them-selves with no help from others."[21]

Towards the end of one of our conversations about what it is like to work as a cashier at Biway, I asked Heather Dixon about her ambitions for her daughter Tiffany. "To make enough to live on without relying on assistance of any kind," she said.

"When we were young we were hoping for the great trip around the world," reflected Marilyn Knox, head of the children's project at the Premier's Council and a senior vice-president of Nestlé Canada. "Their hopes are such simple things. I just hope we haven't taken their imaginations away."[22]

Perhaps as the baby boom ages there has been a demographic shift in aspirations. But Kingston travel agents report that, good times or bad, there is "no question" that students from the elite institution that is Queen's still maintain a strong demand for expensive, long-haul air tickets.

4

THE TRAINING GOSPEL AND THE ARMY OF SERVANTS

> If, instead of a little smattering of Latin; which the children of
> the common people are sometimes taught . . . and which can
> scarce ever be any use to them: they were instructed in the ele-
> mentary parts of geometry and mechanicks, the literary education
> of this rank of people would perhaps be as complete as it can be.
>
> – Adam Smith, *The Wealth of Nations*, 1776

I N THE SECOND YEAR of this century a bitter controversy
erupted at an education conference in Kingston, Ontario. It
became known as the High School Debate. Traditionalists and
reformers squared off in a discussion of curriculum sponsored by
Queen's University and the local Board of Education. At stake was
the issue of whether the traditional academic curriculum should be
leavened with more "practical" training. The principal of Kingston
Collegiate, W.S. Ellis, made a forceful argument that schools should
not simply feed students into the universities. He thought schools
should train personnel for emerging jobs in industry.

Oddly, the questions asked in 1901—what should be taught? to whom?—are still being posed today, during a time when education budgets are shrinking even though the demand for learning is rising. *Today's* High School Debate (one that's spilled over into primary schools and universities) is again centred on whether students should take a wide-ranging menu of liberal arts courses or focus more narrowly on studies that will prepare them for the specific demands of a highly competitive job market. Do schools have a social responsibility? Or should they concentrate on the supply side of the labour market equation? How do education and training fit into a split-level society in which some jobs require sophisticated skills and others demand few skills at all?

In the first-decade version of this debate one of the main pugilists was Nathan Dupuis, a Queen's University astronomer and sometime school reformer. "The liberal in education," he said in 1902, "is an iconoclast"—something he clearly thought a good thing. He called his opponents on the other side of the ring, the conservative educationists, "moss-grown and effete." They were "a class of critics and pedants [who] opposed and retarded the incoming of the age of modern science."[1]

One of those "hoary" opponents was John Watson, a Christian idealist and prominent philosopher of the day. For Watson, a product of the Victorian era of vigorous self-improvement, all the talk of reform and progress and liberalism endangered the old compulsory classical curriculum. The approach would mean overspecialization, a division of the population into those who would enter either the professions or industry and "the counting-house" on the one hand and those intended to be "artizans" on the other. Watson said he was prepared to be labelled "unprogressive and reactionary," but he worried that the move to more "practical" training would encourage rather than moderate the tendencies to overspecialization. It would be better, he said, to cease such "hazardous experiments."

"What we need," Watson concluded, "is an education of the highest kind, open to all; an education which will do the best that can be done for clever boys, however poor their parents might be."[2]

Dupuis responded to Watson in typically pragmatic fashion. He recognized that most children of the day got little formal education and that a high-school education was an expensive proposition—particularly for working-class parents. But this was simply the natural order of things. The unlettered masses would always be with us. "For, after all," he argued, "mankind consists of a few leaders and an army of servants."[3] Dupuis represented a reforming zeal that was concerned with efficiency, not equality. Reform, apparently, had its limits.

In the same decade that the High School Debate was raging in the halls of Queen's, Canada's factory owners were agitating in Ottawa's corridors of power. The businessmen of the day were also seeking to reform the education system. Their eyes were focused not on the halls of learning, but on the new factory floors of industry.

"Our educational systems are devoted almost entirely to preparing pupils for commercial or professional careers," complained George Howell of the Canadian Manufacturers' Association in 1910. "Very little effort is made to interest the pupil, who . . . may desire to work with his hands and through a different system could easily be interested in studies which would tend to guide those hands."[4]

The Canadian Manufacturers' Association wanted more state aid to training for industry. A "new economy" was unfolding even then. The skills of the farm had traditionally been passed from mother to daughter, father to son, in an informal manner, and young people had become multiskilled, learning to be cooks, seamstresses, nurses, teamsters, harness-makers, and horticulturalists, all at once. Any division of labour tended to be along the lines of sex. Similarly, in the early days, goods were for the most part produced in small workshops in which young people were often formally apprenticed, sometimes bound to the same master for years, sometimes striking out as journeymen into the world of independent wage labour. Experienced operatives personally passed on craft skills such as printing and cabinet-making.

By the turn of the century conditions were rapidly changing. Capitalists were establishing a centralized factory system, purchasing machines driven by steam (and soon electricity) and hiring people to run them. The early use of the label "hands" to describe industrial workers says much about the value that management placed on the brains of their employees. The new regime narrowed the division of labour so that different "hands" did different—and smaller—jobs. Owners began to push for two new kinds of trained employees.

First, they required engineers skilled in the design and management of factory systems and machines: modern, practical men instructed by modern, practical men like Nathan Dupuis who appreciated the virtues of industrial efficiency. They also needed new layers of sub-bosses to direct their operations. "The greatest difficulty manufacturers have to face," the CMA said, "is the securing of competent, well-trained mechanical experts to act as foremen, superintendents, managers, etc."[5]

But, second, they also needed a mass of workers to tend their machines, people steeped with a sense of duty and work-discipline, people who brought with them to work what English writer E.P. Thompson has described as "moral machinery."[6] Today's employers have similar goals in mind when, through various team and total quality schemes, they seek to convince workers that the competitive goals of the enterprise should be paramount. Such programs, often labelled as training, are essentially top-down attitude adjustment efforts.

The idea of an industrial division of labour went back at least as far as the eighteenth-century Scottish philosopher Adam Smith, who began his book *The Wealth of Nations* with a classic discussion of the production of pins. The manufacture of pins, said this father of modern capitalism, could be divided into "about eighteen distinct operations" such as wire drawing, straightening, cutting, sharpening, and grinding, "all performed by distinct hands."[7]

Over a century and a quarter later these "distinct hands" required some guidance, it seemed, and in 1904 the Canadian Manufacturers' Association began pressuring the federal government to set up a national training system run out of Ottawa. Their lobbying met with little success until Prime Minister Wilfrid Laurier drafted a bright young civil servant named William Lyon Mackenzie King into his cabinet as labour minister, thus placing the department under its own minister for the first time.

King, a deputy minister since 1900, was well-acquainted with the needs of industry. But since education was defined as a provincial concern under the British North America Act, he knew he was on tricky constitutional ground in trying to deal with the CMA's demands. King, already politically adept, did the time-honoured Canadian thing. He persuaded a reluctant cabinet to appoint a royal commission (the Royal Commission on Industrial Training and Technical Education) to study the matter—but not before he had written each of the nine provincial governments to make sure they would not regard the commission as an infringement on their jealously guarded educational prerogatives.

It was the future prime minister's first solid success as an elected politician. Having spent time in Britain and the United States, he was aware of what was happening in technical education in other industrialized countries. In his capacity as deputy minister he had listened closely to the voices of business. "I am pressing hard [in cabinet] for the Comm'n," he told his diary within days after joining the government, "and see the need for it more & more strongly as I read up on the matter, see what other countries have done & how far we are behind."[8]

One of the other countries visited by King's royal commissioners was Germany, and they had been impressed by Otto von Bismarck's activist state and its involvement in the area of worker training. "Canada is behind the times," they warned. What was needed was more training and less education—something not so "bookish." A new training regime would prepare the coming generations of factory hands with "good habits . . . of obedience, courtesy, diligence and thoroughness." It would persuade people to "accept and fill their place[s] in organized society which implies relative positions and relative degrees of authority."[9]

The Commission also recognized that business owners needed entire new layers of bosses of their own. It urged the government to provide courses of "a suitable sort . . . for highly skilled foremen and managers . . . according to the needs of the particular industry or locality."[10] It also pushed for a rationalization of the existing education system, so that working-class children would no longer be as exposed to studies (especially the "classics") designed for the learned professions. Instead, they should be instructed by teachers with "practical experience," people able to impart whatever skills they might need for the new world of work and at the same time adjust their expectations to produce a class of contented industrial subordinates.

As they travelled the country investigating Canada's industrial revolution and its new labour needs, the commissioners were confronted with the factories that had sprouted up behind the protective tariff walls of John A. Macdonald's National Policy. A stove works in Sackville, N.B., was supplying western Canada. A Charlottetown plant was shipping over half its production of small gas engines to points west of Winnipeg. But Canada's young industrial revolution was concentrated at the centre, not just in Montreal and Toronto and Hamilton but also in places like Valleyfield, Peterborough, Kingston, and especially Windsor, where the truly monstrous factories of Henry Ford had just been established.

Factory jobs were plentiful for new immigrants, city dwellers, and those who were having to leave a farm life on the cusp of mechanization. There was a seller's market for labour. The Locomotive Company at Kingston, where the ticking of clocks and the shrill screech of factory whistles had already replaced the quieter rhythms of smaller-scale production, complained of being "crippled" by worker scarcity. "No good mechanic need ever fear being out of employment," reported an alarmed Commission.[11]

The report of the Royal Commission—published in 1913—reflected a notion dominant among the privileged classes. The masses really had no need for most of the academic subjects that dominated the curriculums of the school systems. Far better that they set their sights on subjects more suited to their stations in life and their destinies in "the manual occupations." Schools should directly address the need of "inculcation . . . of a love of productive, constructive and conserving labour." It was all very well to have the children of "the learned professions" study Latin. Pre-vocational work, manual training, drawing, and (for girls) domestic and household science were not just appropriate, but, indeed, necessary for the "common people" to cope with the new world of modern industry.

The 1913 report was similarly forthright in its assessment of how the growing centralization of goods production was affecting workers. Specialization and the organization of workers in factories had led to a situation, the commissioners stated, in which the "workman" occupied "only the place of a skilled attendant upon a machine," and "only a few individuals are required to understand the machine and have knowledge of all its parts and ability to correct anything that goes wrong with it."[12]

Running the show in the new factories was an expanding class of engineers and technicians—the high priests of the new sciences of production. The factories they designed made people work faster and narrowed what workers needed to know to perform the required tasks. The machines on the line at Windsor's Ford plant or Toronto's Massey-Harris works (the largest tractor plant in the Empire) were "rapid, regular, precise, tireless."[13]

The factory system threatened those who jealously guarded the mysteries of their trades—the workers who had power on the job because they had, over time, learned the skills of cutting metal, cloth, meat, and wood, of themselves fashioning entire machines. According to historian David Landes, their theoretical knowledge was "striking. . . . They were not, on the whole, the unlettered tinkerers of historical mythology." An "ordinary millwright" was often a "fair" arithmetician or mathematician as well, with a knowledge of geometry, levelling, and mensuration. A millwright "could calculate the velocities, strength and power of machines."[14]

Today's most successful machines are no less rapid, regular, precise, and tireless than the ones that Landes describes. Automated teller machines count money out quickly, seldom make a mistake, and certainly never tire—they are on the job twenty-four hours a day. But today's machines, unlike the equipment of the first industrial revolution, do not permit the firms that deploy them to simply break down the jobs of employees into little pieces and make workers easier to train. Rather, they have the customer do the work, which allows the boss to do away with the worker altogether. The Royal Bank of Canada, for example, decided that it would eliminate 4,100 jobs (7.7 per cent of its workforce) in 1994.[15]

———————

When he dropped out of high school Doug Tousignant was fifteen. One of eight boys, he had to go to work when his father got sick and could no longer work on the lake boats. Two of the Tousignant boys were working at the locomotive works on the Kingston waterfront, but Doug headed for the largest employer in town. In 1948 he started as a canteen helper at Alcan. By the time he retired forty years later, Tousignant was a skilled die corrector.

In describing his movement up the skill ladder, Tousignant uses words like "graduate" and phrases like "line of progression." It took years to become familiar with all the tricks involved in working with hot metal. From the canteen he moved into the aluminum plant itself, working as a rod puller and lathe operator before moving into the die shop, where he worked first as a die polisher and then as a die assembler before moving up to be a die corrector. In those years of fabricating aluminum Tousignant and his fellow workers puzzled out two especially tricky custom jobs that he remembers with partic- ular fondness: a façade for Kingston's Hotel Dieu Hospital and the French pavilion at Montreal's Expo '67.

Alcan had no actual apprenticeship program. Instead, like gener- ations of workers who preceded him, Tousignant was exposed to an informal working-class curriculum that blended instruction from his fellow workers with a lot of self-teaching. There was little manage- ment supervision of this process. "You weren't really assigned to a trainer," he says. "You just learned." The problem for him at first was, he admits, that he was not the fastest learner because he was scared—of the job, of the machinery, of failing to get it right. "So one of the guys came down—I was sitting there pretty frustrated—and he said, 'All that machine is going to do is what you make it do. And God damn it, we've told you what that machine will do. Now do it!'"

Tousignant remembers the incident with appreciation rather than resentment—"the best advice I ever had in there." And after he had mastered various jobs in the years that followed, he would pass on the same advice to others—although as a soft-spoken guy he probably wasn't so abrupt. He told learners to be patient, to keep trying, and not to forget that they were in control of that particular operation.

You learned with your own wits and by taking advantage of the pool of skill around you. Older, more experienced workers were always available. They taught him the proper ways of doing things, like how to lift without hurting his back. Sometimes, when the task was more complex, a veteran operator would stick with him for a few days. But this kind of training was by no means easy; the classroom was noisy, hot, and dirty. It could also be dangerous. At one point Tousignant moved into a new job position vacated when the worker who had been doing it was killed.

This unofficial, "each one/teach one" training in which people learned by doing—and often learned by their mistakes—was extensive and continual. It accentuated the natural camaraderie of the workplace, the mutual co-operation reinforced by the fact that, in spite of working for a large corporation, employees identified with each other's efforts.

The postwar years during which Doug Tousignant made his way up the skills ladder at Alcan are a period that has taken on metaphorical associations with prosperity. The very mention of "The Fifties" seems to call forth images of happy families moving into fresh, spacious suburbs, with households jam-packed with shiny new commodities, a bright-coloured, streamlined car (or two) in every driveway. Television brought consumer culture alive; people suddenly had an electronic salesman in their living rooms. At the same time, a labour shortage enabled working-class people to get decent wages. The postwar accommodation, the "deal" between capital and labour, made it seem as if a wet blanket of normalcy had been at last tossed over the flames of industrial conflict. The middle class seemed to be expanding to include most of the population.

Scant attention was paid to worker training, and apprenticeship was something that crusty and slightly old-fashioned voices might promote. *The Globe and Mail*, anticipating postwar wartime labour shortages, called for a revived system of apprenticeship. An editorial in the paper just after Christmas 1945 harkened back fondly to a time when "it was considered quite an honour to be taken on by a leading farmer." The editorialist was clearly cross with those who felt youth was "a time for pleasure" and businesses that took "the short view" by getting rid of apprentices when times were tough.

Formal vocational schools as places of skill training were not the way to go. According to *The Globe and Mail*, "Schools cannot inculcate the loyalty to standards of quality, the respect for the work, which come from personal association with older men whose skilled hands and fund of experience alone can awaken in youth the ideals of craftsmanship."[16]

Of course, students from working-class backgrounds were routinely being shunted into their appointed slots. As the formal school system expanded to accommodate the baby boom, education began functioning as a sorting device, with vocational and commercial classes siphoning off those destined for the factory, construction site, and office. Whether there was room for either craftsmanship or its ideals in such places was uncertain.

What was certain was that business was indeed taking the short view. By 1953 a survey of 709 factories in three industry groupings revealed that only 12 per cent of the plants in iron and steel had organized training programs. There were training programs in 8.4 per cent of electrical apparatus plants, and in 17 per cent of plants building transportation products. This meant that a tiny percentage of the workers in industry as a whole were receiving formal training at work: just over 1 per cent of workers in iron and steel, 0.6 per cent in transportation products, and 1.5 per cent in electrical apparatus.[17]

Industry clearly had little interest in building or even contributing to formal programs that would develop the skills of its workers.

———————————

It could be said that the modern engineer is the descendant of the skilled craft worker. Writer Harry Braverman has shown that the ties between the working population and science were almost completely frayed by engineering and managerialism. Braverman (whose own study of the degradation of work stands as a modern classic) offers the testimony of Elton Mayo to prove the point. Mayo, a Harvard business school and human relations expert whose writings on the social relations of production were to become required reading for student engineers, wrote that in the past: "Someone, some skilled worker, has in a reflective moment attempted to make explicit the assumptions that are implicit in the skill itself. . . . Science is rooted deep in skill and can only expand by the experimental and systematic development of an achieved skill. The successful sciences consequently are all of humble origin."[18]

Engineers and owners sought to liberate themselves from dependence on the skills of craft workers and attempted to reduce and control such skills. In the meantime, owners were dependent upon skilled workers and as a result faced the problem of developing new sources of the labour hierarchies they needed: some skilled workers to "correct anything that goes wrong" and a large mass of much less skilled machine-tenders inculcated with the virtues of "obedience" and a "love of labour."

Such challenges soon gave rise to a whole new "science" (of which Nathan Dupuis may have been only dimly aware): the science of management. As they began to forge an identity of their own, managers categorized the workforce into new compartments: skilled, semiskilled, and unskilled. They also had to come to terms with the erosion of past forms of apprenticeship and on-the-job learning. How to redesign the educational system to produce the skilled workers who *were* needed? At the same time, how to foster a "willingness

to accept and fill one's place in organized society which implies relative . . . degrees of authority"?[19]

For its part, labour was cautiously ambivalent about the issue of
training. Many unionists viewed practical training in schools with
suspicion, as a breeding ground for "rats and scabs."[20] Union stal-
wart Jimmy Simpson (a three-time vice-president of the Trades and
Labour Congress of Canada) did accept an appointment as the
labour representative on the Royal Commission, but was immediately excommunicated from the Socialist Party of Canada for playing
the bosses' game.

The organized worker was wary of technical education if it meant
that it was going to provide a source of strikebreakers or, more generally, if it would hurt them by "flooding the market with workers,
thus reducing wages." Herbert Benson, president of Vancouver's
Trades and Labour Council, supported training, but with reserva-
tions, and told the Commission (as unionists in the 1990s would also
argue) that if an employer wanted the government to supply it with
trained labour, "He should be willing to do his own part." Benson
was also concerned about the growing fragmentation of work. Young
workers should have the chance to learn the "whole trade, from the
bottom up, instead of being made a specialist in one line."[21]

In the 1910s, technical and vocational education was a field as yet
largely untilled—and for the most part unexplored—in Canada.
And although groups like the CMA talked a good line about their own
needs, business was not prepared to pay for training. Provinces were
also not equipped to leap into the field. At the time they had only
restricted sources of tax revenue; income and sales taxes were
unheard of; and, aside from transfer payments from Ottawa, provinces relied on liquidating their timber and mineral resources (often
at fire-sale prices) to fulfil their constitutional responsibilities for
roads, bridges, hospitals—and schools. It was a lean, low-tax, low-
expenditure era.

By the late nineteenth century, educational reformers had successfully promoted compulsory primary education, having persuaded the propertied classes that the modest cost (teachers were

81

abysmally paid) was worth it; the presence on the streets of the idle children of the poor was cause for alarm. Many educationalists were moral reformers who felt that a strong dose of deference to authority could be nothing but a fine thing. As an owner of a factory near Kingston put it, "It is surprising how many men can neither read nor write. Such men are suspicious, and likely to be trouble-makers, so we avoid them. The prevalence of illiteracy emphasises the need for the strictest enforcement of the compulsory education law."[22]

Mackenzie King himself represented a new class of university-trained "experts" who urged more state action aimed at smoothing the edges of a fast-growing industrial capitalism and, not incidently, at making private enterprise function more efficiently.

The first sustained efforts at enforcing compulsory education accompanied generalized reform and expansion of education. Between 1911 and 1931 the average time at school spent per student rose from 6.58 years to 8.55 years; still, by the time of the Great Depression, only 46 per cent of sixteen-year-olds were staying in school.[23]

Compulsory education for younger children was one thing, a national system of technical training quite another. This is what King's Commission advocated: a comprehensive system of training boards, councils, and commissions. It recommended that Ottawa spend $30 million (no small sum in 1913) over ten years to enable the provinces to establish training courses and hire technical instructors.

But by the time the Commission reported in 1913 the ruling Conservatives under Robert Borden were occupied elsewhere. Recession hit that year and World War I soon followed. It was not until the eventful year of 1919 that Ottawa finally turned its half-hearted attention to national measures aimed at "training." The Bolshevik Revolution was sweeping aside the old Russian Empire. Canadian workers capped an unprecedented wave of union organizing and demands for the eight-hour day with a series of general strikes that challenged the existing order. The summer of 1919 saw 115,000 workers walk out in a total of 210 strikes.[24]

The authorities responded to the upstart proletarians with clubs, guns, and prison sentences. Ottawa also launched another royal commission (this hasty version dealt with industrial relations and had a report ready in five short months) and brought together some of the country's most moderate unionists to talk with management.

Ottawa's reaction to the labour strife also included the Technical Education Act of 1919. It took some cues from Mackenzie King's Royal Commission, but there was to be no comprehensive system of local boards. The plan introduced a shared-cost program under which Ottawa was prepared to invest $10 million over ten years, *provided* the provinces came up with matching funds. The Commission had proposed three times this level of spending and called for a more generous expenditure of federal dollars, proposing to pay 75 per cent of the cost of worker training. Finally, the act was aimed specifically at promoting vocational education in secondary schools rather than training for workers who had left school.

Astonishingly, it was not until 1948 that the provinces took up the last of the $10 million offered under the 1919 act. Such was the disinterest of the provinces and their inability to match the federal offering that the original ten-year plan had to be extended four times.[25]

The Canadian response to the oft-repeated need for a skilled workforce was essentially one of laissez-faire. Employers might call for training, but they were always reluctant to do much about it themselves. Government followed the advice of practical men like Nathan Dupuis, so that even though the education system slowly took up vocational training, it did little more than frog-march the children of the working class into their appointed places in workshops and, as the years passed, in offices and stores. The sort of thinking represented by John Watson and his more influential U.S. counterpart John Dewey (who insisted that we should not be making children suitable for industry, but industry good enough for children) was slow to gain influence.

During the interwar years there was a slow growth in vocationalism (or "tech-voc") in the high schools. It was still controversial to raise the issue of regulating child labour, and although more young people were getting a few more years of education, most still failed to get far into high school—let alone pursue any comprehensive course of vocational training. The idea of skill training tended to be set aside. The reorganization of work was left to employers who adhered to the principles of scientific management, under which they made every effort not to enhance the skills of workers, but rather to *deskill* the labour process.

Technical or vocational education was larded with class implications, which its boosters did little to gloss over. Whether it was Nathan Dupuis's "army of servants" or the training commission's good habits of "obedience and diligence"—or even its recognition of mass-production workers as little better than slaves—it was clear to anyone who cared to listen that the vocabulary of skill training (not to mention its methodology) reeked of hierarchy. Those who spoke of training were candid enough about who stood where in this hierarchy, whose vocation it was to serve. Those who lacked the means of access to education and had no craft skill were clearly at the bottom. The situation was at least marked by candour, not by a legitimizing veneer of "empowerment" and "life-long learning."

There were "at least two castes" among the populace, said Albert Leake, Director of Technical Education for Ontario: "Those who are the elect and those who are not, i.e. those who can absorb the printed page and pass the prescribed examinations and those who for mental or financial reasons are not able to do so."[26] This sort of unvarnished talk is refreshing in its candour, particularly in comparison with the glossy veneer of "lifelong learning" and "learning a living" that would come to dominate public discussion of worker training in the 1990s—when opportunities for good jobs are in fact fewer than they were in the immediate postwar period.

The enthusiasm for workers education expressed by the Canadian Manufacturers' Association at the turn of the century had evaporated when the businessmen saw that government was willing to do two things: import skilled labour and pay for rudimentary vocational training in the formal education system. Big manufacturers were also wedded to scientific management, whose goal was less dependence on skilled labour. When Ontario proposed, for instance, that the crucial trade of tool- and die-making be officially recognized under the laws governing apprenticeship—which would mean formal instruction in the province's big industrial plants—the CMA nixed the idea.[27]

Not all employers took the CMA's position. A notable exception— and an outspoken advocate of more management action on training—was Joseph Pigott, who ran a big construction firm in Hamilton. Pigott (a long-time member of government training advisory councils) denounced the "apathy of employers." He felt their participation in skill development was "absolutely essential," and that bosses needed bosses. "For some years in Canada it has just not been possible to pick up foremen, superintendents and leaders," he said, even though since the 1920s, "Immigration from the British Isles provided a ready supply of mechanics."

In 1953 Pigott summed up changes in his own business and presaged the debates of the 1990s over the needs for particular kinds of workers:

> There are some people who . . . feel that buildings will go up in any event—if not out of the traditional materials built in by the skilled trades, then by aluminum, glass, stainless steel, factory assembled units. . . . They apparently assume that a country does not need skilled men; machines will replace them. They forget that a journeyman carpenter, for instance, has a trained mechanical mind and he has expert fingers; that he is a real asset to his country because he has flexibility."[28]

Both government and industry treated worker training with benign neglect. The system of apprenticeship had already weakened, and during the postwar period the trend continued. Out of 38,280 workers entering formal programs between 1945 and 1952, 11,640 people quit, a dropout rate of 30 per cent.[29] By 1956 federal funding under the Vocational Training Co-ordination Act barely reached one million dollars for all provinces except Quebec and Prince Edward Island.[30] Correspondence between training officials from Ottawa and Ontario was filled with gloomy phrases such as "a long way from desirable" and "doomed to failure."

By the 1950s formal apprenticeship was generally confined to a few of the building trades, motor vehicle repair, and the barbering/hairdressing sector. High schools offered a grab bag of courses aimed at the vocational stream. Industrial arts and home economics supplemented general programs, but were by no means truly comprehensive vocational courses of study. Rather, they were filters. In the words of a 1961 federal study of technical education, vocational programs "provide exploratory experiences which help students to make realistic occupational choices according to their interests and abilities."[31]

High school had become a common experience for most Canadian teenagers. One scholarly study of Ontario education, fondly recalling the days of duck tails and crew cuts, bobby socks and penny loafers, called it "high school's last great decade." According to educational historian Robert Stamp, the period was "a kind of golden twilight before the radical program alterations and student unrest of the 1960s."

The high-school population began to reflect the population at large, although the upper middle class was still overrepresented among actual graduates. In 1956 an Ottawa high-school administrator reported, "We get proportionately more people of average or poor quality of brains than was the case formerly."[32] Schools specializing in technical and vocational education offered three- and four-year courses, but supplemented these with shorter "terminal" courses designed to equip dropouts with "marketable skills."[33] However, youth with little or no formal qualifications could still find a niche in the job market. There were both unionized positions with decent wages and the usual array of low-wage jobs that could be filled by urban youth as well as people migrating from Canadian farms and southern Europe, particularly Italy.

But that did not address the vexing question of how to produce and how to retain skilled labour. For Canada, a land that had long prided itself on its success as a trading nation, the solution was obvious. Skilled workers, like any other commodity unavailable domestically, could simply be imported.

And imported they were, in their thousands. In the late forties and early fifties the economy expanded and unemployment was low. In earlier years Ottawa had simply left immigration to outfits like the CPR that needed workers. The government subsequently took an active hand in promoting immigration, concentrating on Interior Minister Clifford Sifton's famous eastern European sons of the soil with their "sheepskin coats" and "stout wives." Sifton's moves aside, Canada had always discriminated in favour of "British subjects," and the post-1945 period was no exception. Mackenzie King said that although Canada would encourage immigrants, it had no desire to "make a fundamental alteration in the character of our population" or to change its "fundamental composition."[34] The new department that regulated the flow of immigration clearly favoured British subjects, followed by other Northern Europeans and Italians. Between 1951 and 1957 the largest single category of blue-collar immigrants was skilled and semi-skilled workers from the United Kingdom. Next came less skilled labourers from Italy, followed by more manufacturing workers from Germany. In the 1950s the United Kingdom, Germany, and the Netherlands supplied virtually all of Canada's semi-skilled and skilled working-class immigrants.[35]

"The postwar immigration proved to be largely urban, with skilled labour, business and professional qualifications far more in evidence than among earlier waves of immigrants," political economist Reg Whitaker concludes. "The state was now much more squarely at the centre of the immigration process."[36]

The free market in labour had its downside for the trading nation, which also exported skilled people, mainly to the United States. For every skilled German who arrived in Canada during the 1950s, a Canadian headed south. Nevertheless, the strategy seemed to work. In 1954 the training supervisor for Canadian General Electric in Peterborough told a meeting of Ottawa's apprenticeship advisory committee that the skill shortages of the day would have been far more serious without the tide of immigration. Even though L.J. Sparrow was in charge of apprenticeship at CGE, he reflected the prevailing corporate attitude to active training, arguing that industry was best kept as free as possible from the restrictions around who could do what job. According to the committee's minutes, "Mr. Sparrow felt that in a free country a worker who could learn a trade outside the apprenticeship system should not be prevented from working at a job he can do."[37]

Employers simply found it easier and cheaper to hire pretrained workers than to undertake training themselves. The government commitment to laissez-faire was ambivalent: while it was prepared to regulate the flow of migrants and manipulate immigration content to meet the needs of industry, it was not prepared to do much active training itself. It certainly did not force private business to train.

Like Doug Tousignant, workers would arrive on the job and simply learn by doing. Some would lie about their experience. In many cases the boss just wanted a warm body who could figure things out by the pick-up method. This was before the notion of "workmate" had been appropriated by the Black & Decker company in an attempt to sell stand-alone vices to do-it-yourselfers. The informal working-class solidarity that prevails on the shop floor took the place of more formal training systems. Many simply learned from their workmates. They learned by imitation and informal instruction. It was the system that produced skilled workers.

Nevertheless, the idea of apprenticeship—with its appeal of formal learning while doing—hung on stubbornly. This may be because, like today's co-op programs that combine formal education with on-the-job experience, apprenticeship is an effective way of learning to do particular kinds of work. In Ontario, with its long-established though long-neglected apprenticeship system, provincial officials refused to give up on the idea. But successive directors of apprenticeship came to realize that in the big plants and construction projects that were humming along in the fifties and sixties, the prevailing system was informal training relying on internal labour markets.

In 1952, not long before his retirement after more than twenty years as apprenticeship director, Fred Hawes explained this rough-and-ready system to Ottawa's Director of Training, A.W. Crawford. Employers paid little attention to the "proper selection" of workers: "Some employers just select the first young lad who comes along while others favour relatives and friends." For Hawes, this was an unsatisfactory means of recruiting apprenticeships because the lads who entered the workplace often had "no desire, ability or ambition."[38]

In 1955 Hawes's successor, G.H. Simmons, told Crawford about the "inadequacy of existing efforts to promote and develop suitable training programmes for apprenticeships" in the building trades. His remarks could have been applied to other trades that needed, in his view, years of skill development.

> The fact that employers are not greatly concerned about the existing situation would seem to indicate that formal apprenticeship is not essential . . . and that the prevailing practice of learning by the pick-up method is suitable for the great majority of learners. If, as I believe, this is not the case, it is evident that we must look elsewhere for a satisfactory situation.[39]

Seven years later, Simmons's satisfactory solution had still failed to emerge; most apprentices in Ontario were either barbers or car mechanics. Formal on-the-job learning—which had never taken root in the factory—was even withering in the building trades. The new apprenticeship director, D.C. McNeill, called the entire situation "disgraceful." In a confidential memo to the labour minister, an outraged McNeill pointed out that construction managers were borrowing ideas from the deskilling techniques of scientific management that dominated the factory system. Employers, he said, would hire unskilled and half-skilled people and pay them higher wages than they would pay to an apprentice. "The young men are not receiving any particular training but are simply doing 'jobs', with the result that the trades are being broken down into little pieces. . . . The only trades where there seems to be any effort to maintain the apprenticeship plan to any degree is in the union shops."

McNeill, worried about the quality of production, described a typical building site where one skilled bricklayer was hired to build corners and openings, while a dozen men "who cannot be classed as anything much more than labourers" laid most of the brick.[40]

Aside from the way work was being organized into fragmented bits, there was another fundamental reason for the employers' failure to train. If government was content to import skilled labour, business was just as happy to buy that commodity on the open market. This often took the form of "poaching." Just as hunters steal onto someone else's land in search of deer or duck, employers regularly hire another species—the skilled worker—away from competitors who may or may not have taken the trouble to train that worker. Indeed, one comprehensive dictionary offers a second definition of the verb "to poach" as not something you do to an egg but "to encroach on or usurp or steal (an idea or *employee*, etc.)."

As Ontario's Hawes told Ottawa's Crawford in 1953, a major part of the blame for the problems with training lay at the doorstep of "the unethical employer who in times of good business when good mechanics are hard to obtain does not hesitate to bribe the apprentice away from his employer."[41]

Training advocates from individualistic North America have routinely looked to Germany and Japan for inspiration. There, they say, a sense of mutual obligation—albeit mediated by the state—exists between workers and bosses. German sociologist Wolfgang Streeck has emphasized the "social obligations to train."

> In open labour markets, employers competing with each other will always be under a temptation to 'cheat' by not investing in general training and hiring skilled workers from their competitors.

What is more, the mere prospect that others may behave in this way is likely to deter employers from training even if the result will be a general skill shortage. Societies that have at their disposal institutional or cultural mechanisms by which to *oblige* firms to co-operate in training are likely to enjoy competitive advantages as they will be able to protect their firms from the dysfunctional consequences of market-rational behavior for the production of skills as collective goods.[42]

But the Canadian state was not prepared to oblige firms to train. Nor did employers pay particular attention to training—even in the midst of postwar labour shortages. One employment officer at the Unemployment Insurance Commission reported in 1952 that his Barrie, Ontario, office was getting a lot of inquiries regarding apprenticeship training, but "employers are reluctant to enter into long term agreements because of the uncertainty in work continuity."[43] Part of this uncertainty may well have been tied up with a tight labour market in which workers were in a much stronger position than they would be forty years later, when an official unemployment rate of anything below 10 per cent would be a giant step forward; in 1952 workers were more likely to tell the boss to take his job and shove it. But the uncertainty was also caused by the possibility that other bosses might *poach* workers who had acquired the skills in demand.

Still, employers did not hesitate to complain about the skill shortages that resulted, urging government to get more actively involved in the supply side of the labour market by doing something about worker training. By 1966, when unemployment stood at 3.4 per cent, the CMA claimed that there were thirty thousand vacancies for skilled workers. The association sent a letter to its 3,200 Ontario members urging them to hop on the training bandwagon gathering momentum once again. Skill shortages, according to the employers, were a "problem for all manufacturers," particularly when employment was "running at its current high."

"In the past," the association admitted, "employers have managed to reduce their manpower shortages from the supply of immigrants, as well as [from] a normal recruitment from other companies, but the position has now been reached where demand exceeds supply."[44] Low unemployment is always bad news for ambitious managers. Being self-interested actors in the best tradition of Adam Smith, they view the labour market's supply/demand equation from the buyer's perspective.

So long as labour demand was anywhere nearly balanced with supply, as it was in places like Windsor for much of this period, you could indeed "quit at Ford's in the morning" and get on somewhere else in the afternoon. This was particularly true if you were a machinist, millwright, electrician, or any other skilled worker. Your bargaining power was enhanced by a strong demand for your skills that—combined with low unemployment—made you truly flexible. But even if you had little formal education or training (in 1961 over one in three Windsor wage-earners had never attended high school) your chances of finding work were still good. Those with high school were even better off, as unemployment among workers with elementary school education was nearly double that of the wage-earning population. But the fact remains: the labour market was a seller's market. One of the first research studies published by the new Economic Council of Canada (founded in 1963) examined labour-market conditions in Windsor. It showed that 1964 unemployment figures among workers in several of the most important trades (millwrights, tool- and die-makers, machinists) were too small to report.[45]

Again, this had its effects on management's attitude to training. Border Tool and Die Ltd., a Windsor firm with an active apprenticeship program in the mid-1960s, reported that it paid top dollar to its starting trainees, gave them regular raises, and furnished "proper training." General Manager John Tingle was a member of a government apprenticeship advisory committee. However, by late 1968 the firm had decided to withdraw from any broad apprenticeship program, citing the "cut-throat hiring techniques" of competing companies. Border Tool decided that from then on it would be narrowing its training by tailoring it to the specific needs of its plant: "We are going to try our best to train our boys our way for our plant and not give them extra training and schooling to benefit some other company."[46]

This is a far cry from the sense of wistful corporatism served up by the 1913 Royal Commission—the emphasis on "solidarity . . . citizenship . . . community" that it saw as holding sway in Germany.[47] Today's training gospel has a distinctly different emphasis. Now we hear constantly about the need for workers to equip themselves with the new skills demanded by a competitive globalizing world. Still, though the words may be different, the melody lingers on. The questions—what skills? what workers? what jobs?—are resoundingly familiar.

5

PUBLIC PURPOSE, KNOW-HOW, AND THE WAR OF PRODUCTION

Freedom for the individual does not lie in pushing back the increased role of government. If we are concerned that people as a whole, not privileged groups, should truly enjoy more freedom, we have to make more collective provision to protect people against poverty-inducing disasters and to make positive opportunities accessible to all.

– Tom Kent, *A Public Purpose: An Experience of Liberal Opposition and Canadian Government,* 1988

I N 1966 a frustrated Margaret Lett wrote to her representative in the provincial government in far-off Toronto. She said she was commuting every day in an aging car from Eganville, Ontario, to the nearby Ottawa Valley community of Pembroke, where she worked as a saleswoman at Woolworth's. She wanted to see if she could take courses to qualify as a hairdresser. That way, maybe, she could improve her job position, perhaps set up a hairdressing shop on her own and avoid the daily drive.

But, she wrote, the approved three-year apprenticeship in hairdressing was both too long and too expensive for her to consider. She had her eyes set on a nine-month course at the private Marvel School in Ottawa. How could the government help?

"I am a separated 28-year old and have two school-age children to support," she said in the letter. "My husband is someplace in B.C. as far as I know and he hasn't been giving me money as he was ordered. With the rent to pay, transportation, the welfare of my children etc I find it almost impossible." If she was able to take the course in Ottawa, she added, her parents would be able to care for her children. The three-year program was simply too long, she said, because she wanted to have more time with her children.

The provincial government was in the middle of an explosion in growth and complexity, but there were not yet that many layers of administration and Lett's appeal found its way to the desk of the deputy minister of labour, whose department was responsible for worker training. From there it filtered down to a labour department counsellor, who duly arranged to meet Lett at her Eganville home. The appointment fell through when Lett's car broke down in Pembroke, but the man from the ministry did manage to contact her by phone. He told her there was nothing to be done, and that if she wanted to become a certified hairdresser she would have to take the full, three-year course.[1]

Margaret Lett's story will be familiar to many working-class women, with or without husbands, who have faced the need to adapt to the demands of the formal job market and at the same time have to balance that need with the complex job of "housewife." The question is, where to turn for formal training, particularly during a time when women were almost always slotted into low-level work in the service sector?

Of course, the notion of the full-time housewife has scarcely ever applied to all women. Although (as the Royal Commission on Training pointed out in 1913) "the great fundamental occupations" of housekeeping and homemaking were "still almost exclusively in the hands of women" (for which they had a "natural liking"), many girls left school to go immediately to jobs in light industry and stores, where they worked on knitting and sewing machines, stocking shelves, taking inventory, or making up invoices. The road to corporate concentration was already clearly discernible, with the individual shopkeeper being replaced by "the huge emporium which employs a large number of sales clerks. With each clerk confined in *his* activities to one department, the requirements of knowledge . . . while not less exacting, are less complex and comprehensive than formerly."[2]

With scientific management and fragmentation well under way, modernizing reformers not surprisingly decided that young women should be given a chance to develop "vocational ability for housekeeping." Even in the area of "domestic science" some were worried that Canada was falling behind the competition. As usual, Germany was held up as the shining light of vocationalism, with "the Kingdom of Prussia alone" having fifty Stationary Housekeeping Schools and forty-one Itinerant Housekeeping Schools. The Commission's recommendations included elaborate schemes for the training of houseworkers by resident or travelling "instructresses" and in "middle housekeeping schools." One area where the authorities heeded the Commission's urgings was in home economics; the Commission argued for housekeeping courses for younger girls and for schools for domestic science. There was even provision for young women to learn the principles of "sanitary administration."[3]

Yet girls had been learning domestic skills from their mothers for centuries; in fact, such on-the-job training was the kind of tacit apprenticeship that predated the way many male industrial workers picked up *their* skills. People have always learned—not from their bosses or in formal training settings—but by watching and learning from each other.

Girls cared for their younger siblings and started to sew and cook and clean without the benefit of "domestic science." Working-class women and children worked in the home to add to the family income and make ends meet. Women took in boarders and laundry.

They sewed at home on a piecework basis. They worked as char-women in the homes of those members of the middle class who could not afford permanent domestic servants. They kept small shops and baked bread, pies, and cakes for sale in an informal sector that existed long before today's preoccupation with an "underground economy." When women and their children did this kind of work it usually went unrecognized in the same way that ordinary housework was hidden, escaping the attention of statisticians.[4]

In the period after World War II the shift in the structure of the economy gained momentum. Although many married women stayed at home to fulfil all the other functions of housework, partic-ularly child care, the service sector was slowly becoming the domi-nant employer. This meant changing work for women. They had once spent a lot of time baking and making the children's clothes. But the introduction of mass, factory-produced "ready-mades" fore-told an era when some housework could be replaced by industrial products like bread and the clothing advertised and sold in mass chain stores. As time passed, the factories producing the ready-mades either moved offshore or needed less labour, and most of the new jobs were created in white-collar and pink-collar service work. The trend would lead to work in fast food and in advertising fast food. The charwoman would be at least partly eclipsed by commer-cial firms with evocative names (Molly Maid and Maid For You).

C. Wright Mills's spectre of society as a great sales room and enor-mous file was coming clearly into focus. "Fewer individuals manipu-late *things*, more handle *people* and *symbols*," Mills wrote in 1951. A growing army of workers was living off "the social machineries that organize and co-ordinate the people who do make things."[5] For the women in the new labour market, a parallel process was at work. The service sector's takeoff was based on a supply of cheap female labour. At the same time the gradual but accelerating use of women's wage labour outside the home pushed demand for replacement services like fast food and, eventually, day care.

The formal education system lagged behind the changes in the economy. Before 1960 Canada offered little postsecondary education in the technical and vocational (or "tech/voc") fields, with less than ten thousand students enroled in a variety of courses from horticulture and land surveying to metalworking and electronics. Of the almost 7,500 students in such courses in 1959, 38 per cent were taking the male-dominated electrical and metal classes. In fact, the whole area of formal skills training was essentially a male preserve; only 5 per cent of the total student population was female. Few women were registered in engineering and scientific courses.

This was simply a duplication of the existing high-school structure. In 1959 two out of every three high-school students in the vocational sector were taking commercial or home economics courses, provided primarily for young women. According to a 1961 survey of technical training conducted by the federal labour department, "The large enrolment in commercial courses reflects the continuous demand for office workers . . . a field which is particularly suited to the employment plans of most young women."[6]

By 1961 the "employment plans" of young women were changing quickly along with the frenetic growth of service-sector jobs. Sales and secretarial work was plentiful in the fresh new shopping centres and high-rise office towers. Home economics and commercial courses were designed to prepare teenage girls for jobs in offices, stores, and homes. What's more, the country was on the edge of a huge upsurge in nursing and teaching jobs that accompanied the growth of health care, social work, and education, the welfare state "industries." But implicit in the prevailing conception of women's employment plans was an ideology of domesticity, in which *home* economics meshed smoothly with the world of paid work, a world that women would enter and leave according to the demands of their primary job—that of housewife.

Many young, working-class women were untrained for the world of white- and pink-collar work. Some turned for the skills they needed to private technical schools, big and small, that dotted the urban landscape. The Ontario education ministry files on the private training schools reveal an industry both widespread and largely unregulated.

The Trade Schools Branch of the ministry issued licences to people as "Salesman of Trade School Courses." A one-dollar registration fee enabled operators to advertise themselves, as the Gravenhurst Business College did in 1952, as "Fully Accredited by the Department of Education, Province of Ontario." F.W. Ward, the civil servant responsible for private-sector training, told one of his licensees, "This Branch operates with a maximum staff of two (including me)," and he went on to describe this state of affairs as "adequate for all months of the year" except for a short period before the licence renewal deadline. Mr. Ward's files reveal no record of inspection.7

The private training business was conducted by a handful of big chain operations (some headquartered in the United States) and a large number of independently owned local operations. The mom-and-pop segment of the business was, interestingly, dominated by mom. Of 105 active ministry files for trade schools for 1952, 25 per cent were small businesses owned and operated by women—a remarkable example of independent activity by women at a time when, according to popular mythology, they were supposed to be at home with the kids. Most of these women were training other women in office skills such as typing, shorthand, and basic book-keeping, although one proprietor described her occupation as "housewife" on her registration form. Most of the schools were small, like the Drummond Business College in Renfrew, which offered day and evening courses to sixty-one people in 1952.

At the other end of the scale was the Chicago Vocational Training Corporation—"Chartered as an Industrial Training Organization Training Ambitious Men for Better-Paying Employment" in diesel, automotive, refrigeration, and maintenance mechanics. Then too there was the chain of Marvel schools—"The Largest and Most Modern System of Hair and Beauty Culture Schools in America"—whose sales efforts had attracted the attention of single mother Margaret Lett. Well before Lett wrote to the government seeking help with her training problem, the labour department's apprenticeship director told his boss, deputy minister J.B. Metzler, that the Marvel people were "rather good salesmen" who once ran barber colleges that "became a money-making scheme for the owners and flooded the market with half-baked barbers." In 1961 the province had to keep a sharp eye on Marvel's operations in Ottawa, where they had no competition in the field. "They are one of the greatest headaches we have in the Apprenticeship Branch," D.C. McNeill reported. "One of their main concerns is to get as many people through as possible in the day."[8]

By that time government in Canada had started to move on the issue of training. The Diefenbaker Tories brought in a Technical and Vocational Training Assistance Act (TVTA) in 1960, the biggest government expenditure on training Canada had ever seen. At the time unemployment was relatively high—it hit 8 per cent that summer—but employers and government noticed that there was also a significantly high number of job vacancies. The traditional Keynesian device of stimulating demand would not work in this situation. The TVTA marked the start of a flurry of training and education initiatives aimed at the supply side of the labour market, at turning out workers with the skills needed in an emerging "postindustrial" economy. Between 1919 and 1960 Ottawa had dispensed about $100 million on capital projects, training allowances, and operational training. Under the TVTA, which lasted from 1960 to 1966, the federal government spent $259 million on operating costs *alone.* By the time the program ended in 1967 it had dispersed $592 million on capital funds for buildings.[9]

In spite of the impressive sums spent under the TVTA, it was similar to the Technical Education Act of 1919 in that it provided federal dollars that provinces had to match. The predictable result was that the richer provinces (especially Ontario) took the program and ran with it, using it to subsidize their school systems. Ontario, with the lowest rate of unemployment in the country, was absolutely dexterous at pulling in the federal money, getting the bucks, exhibiting "remarkable skill in devising means of exploiting every letter of the legislation." In less than three years it had used up nearly $200 million in capital funds. Quebec had received $28 million.[10]

Although education was a jealously guarded provincial responsibility, Ottawa found itself, through the TVTA, controlling salaries paid to maintenance workers in vocational high schools. Joey Small-wood's Newfoundland government used TVTA money to underwrite his cabinet's private dining room—the chef was paid as an instructor at the vocational training institute. What's more, although the TVTA had a provision for upgrading the skills of employed and unemployed workers already in the labour market, it failed to reach the people most in need of help—older workers with inadequate formal education. The average age for trainees under the program aimed at the unemployed was twenty-three.[11]

Industry "not only accepted, but demanded" government spending on on-the-job training (OJT). In one case, in the textile industry, the money must have seemed like manna from heaven. One TVTA project aimed at upgrading the skills of women who stitched garment cuttings together paid the trainees the minimum wage of a dollar an hour, with the cost split evenly between employer and government. This was administered by company supervisors, the workers were classified as "semi-skilled," their work typifying the segmented division of labour that had come to characterize many factory jobs since the turn of the century: "Work is broken down into a large number of machine operations. . . . Sewing machine jobs require the ability to do routine work extremely rapidly." According to a project report, "Production returns accrued 100 per cent to the company while training costs, other than wages, were totally absorbed by the government."[12]

This arrangement was too much for the government, which altered the terms of training in subsequent garment industry programs so that employers paid more of the cost. However, a cost analysis revealed a combined federal-provincial subsidy of 44.3 per cent. The workers absorbed 34 per cent of the costs, while the employers paid the remaining 23.7 per cent.[13]

Such initiatives coincided with major welfare state initiatives of the decade, especially on health and education. Saskatchewan introduced medicare, university enrolments increased, and a separate layer of new technical or "community" colleges was created. Welfare and unemployment programs expanded, and the government took up more active labour-market policies. All of these initiatives would create tens of thousands of new jobs in a mushrooming service sector. Nurses, teachers, building maintenance and clerical workers, and social workers would join legions of private-sector service workers—everything from assistant vice-presidents for product development to bank tellers—on the march into new places in the job market.

It was a heady period. "We assumed, accurately enough at the time, that the economy could be managed to offer enough jobs," Tom Kent, Lester Pearson's principal policy advisor during the 1960s, recollected almost thirty years later.[14]

This was not the world that, during the postwar boom, had emerged in the fertile mind of a young science fiction writer named Kurt Vonnegut. Vonnegut's first novel, *Player Piano*, published in 1952, is set in Ilium, New York, a town divided into three distinct districts. A few managers, engineers, civil servants, and other professional people live in the northwest. In the northeast there are the machines. Across the Iroquois River in the south is "where almost all of the people live."

Illium is a world in which skill (or, in the town's all-American vocabulary, "know-how") is the key to victory in the "war of production." Vonnegut describes the mad scramble for places at Ilium's community college, where there are six hundred applicants for twenty-seven openings. It is a world full of anxiety and insecurity: "Paul could see the personnel manager pecking out Bud's job code number on a keyboard, and seconds later having the machine deal him seventy-two cards bearing the names of those who did what Bud did for a living—what Bud's machine now did better."[15] Substitute "human resources manager" for "personnel manager" and "printout" for "cards" and Vonnegut's imaginary world of 1952 becomes the real world of the 1990s.

———

One of the innovations the Pearson Liberals brought to Ottawa was the idea of an economic council, which the reform wing of the

Liberal party hoped would offer advice and help to build a sort of German-style consensus. In its first annual review in 1964, the new Economic Council of Canada called for investment in skilled and technical workers. The 1965 Throne Speech, written by Tom Kent, similarly declared that although the economy seemed to be humming along nicely its "great potentialities" were being neglected. "The talents of some of our people are wasted because of poverty, illness, *inadequate education and training*, inequality in opportunities for work."[16]

The dominant idea of the times, as Leon Muszynski put it years later, was to improve the "marginal productivity" of those many people who were falling short of the mark by "investing in human capital improvements such as education and training."[17] The perspective has remained dominant, scarcely unchanged in the years that followed. It is the scenario of liberal individualism: it places the onus on individual workers—those empty vessels who have to be filled up with skills if they are to survive the rigours of the free market in labour. This is the perspective that prevails in discussions of training that hold that the solution to poverty is a job—and that the way to get a job is to load up on skill.

The approach emphasizes equal access to education and training, recognizing that the young Portuguese woman taking up the 1965 garment-industry training program at minimum wage had not had the same life chance as the young second- or third-generation Canadian man whose parents were socially positioned to provide him with access to university that same year. One way that such imbalances might be addressed in any serious manner was through affordable public education and training. In the 1960s Canadian governments began to make unprecedented investments in schools, colleges, and universities, which—together with student grant and loan programs—for the first time put higher education within the grasp of most Canadians.

When the Pearson Liberals decided in 1965 to try for the majority government that had eluded them in 1963, one of their election pamphlets was called "Canada Needs Trained Minds." And in spite of the left-liberal concern with poverty exhibited by some members of the federal Liberals, the more fundamental reasons for the revival of training as a policy priority can be found in Ontario, the province that was briskly scooping up a disproportionate share of federal TVTA money and pouring it into vocational education and training in industry.

Ontario took one of the first steps to rationalizing its educational system to more closely tailor it to the needs of the labour market with its "Reorganized Programme" of 1962. Dubbed the "Robarts Plan" after the education minister who was seeking the premiership, it included a two-year "occupational" program to prepare for service industries. According to Robarts, the streaming scheme was aimed at preparing young people for "the particular vocations" they would want to enter.[18] Ontario appointed a select committee to study training, hiring John Crispo, a young industrial relations specialist, to do the research. The committee advised that with technological change leading to fewer unskilled and semi-skilled jobs there was an urgent need for retraining with "the on-going shift in favour of the service sectors." According to Crispo, the young people then cascading into high schools had "to plan their educational careers carefully so as to provide themselves with an adequate form of long-range employment security."[19]

At a 1963 conference organized by the Robarts Conservatives on "Automation and Social Change," one of the invited speakers was Sir Geoffrey Vickers, an early futurist whose book *The Undirected Society* looked forward to the arrival of the leisure society. "In the past the leisured class has been drawn from the rich," Vickers told the guests. "In the brave new world which is coming, only the rich, it seems, will be busy—or at least busy of necessity."

Vickers did not ignore the uneven application of the looming leisure era, noting that leisure might well be "compulsory" for some, taking the form of "open or concealed unemployment." Nor did he neglect the possibility that the decreasing need for work might lead to either good or ill. But the former industrial planner, once second in command in Churchill's Office of Economic Warfare, still felt that the dystopian worlds envisioned by Orwell and Vonnegut were not inevitable. He hoped that people could find an "escape from the monkey world of techniques into the truly human world where they can exercise the inalienable activities of human beings, skill, appreciation, sympathy, delight and wonder."

A more sobering note came from David Archer, president of the Ontario Federation of Labour. Archer, who had been reading the work of U.S. management theorist Peter Drucker, mused about the arrival of a "second industrial revolution." Back then Drucker was warning that schools were doing students a disservice by simply preparing them for their *first* jobs. Drucker felt that long-range employment security would soon be a mirage and that under automation it was certain that "even the bottom job" would change "radically and often."

It was left to the trade unionist to interpret the meaning of all this for those who would have jobs in the brave new world that was coming: "The traditional bonds of skill and association that formerly bound workers together in unions will be lessened," Archer predicted. "Basically unions will be endeavouring to protect the jobs or at least the wages of the employees. Management . . . will endeavour to make as few commitments as possible."[20]

———————

The men who worked as stationary engineers in Ontario provide another example of the extent to which workers have relied on their own resources to acquire the skills they needed. In 1968 the department of labour undertook a major survey of these skilled mechanics who maintain large engines, generators, and other machines in industrial plants. The department sampled 1,606 such workers and found that fully two-thirds had learned their trade "by casual means." That is, they learned from each other, outside the context of any formal training system. They were typical of male industrial workers of the time. Some 90 per cent of them had less than Grade 11 and, although six out of ten indicated an interest in training, 86.7 per cent were not active in any formal training at the time of the survey. These were veteran workers; their mean age was forty-eight.

"Just because education is regarded as the panacea for everything today and training sessions are the 'in thing' in occupational and trade groups, does not necessarily mean that they are viable solutions to problems," noted the study. "We can see, for instance, that stationary engineers are not exactly falling apart as a group, even though the majority . . . have limited educations, are over 45 years of age and have picked up their training on the job."

Like many subsequent reports, this one underlined the importance technological change would have on the workforce. Indicating that the trade of stationary engineer was likely to decline in the near future, it concluded, "Industry will . . . encourage workers who are classed as Stationary Engineers to become multi-functional."[21]

In this case "multifunctional" implies a continuation of the process of deskilling familiar to many workers. Employers would deploy new technology and—combining it with the principles of scientific management—force a once-autonomous group of skilled maintenance mechanics to abandon their traditional trades. They would become more "flexible," which means they would do a greater number of less-skilled tasks. At the same time that this kind of flexibility was imposed on workers, employers would themselves gain their own form of flexibility by contracting-out everything from janitorial services to those of electricians.

A 1968 report from the Ontario labour department sheds further light on these imperatives. Using Dominion Bureau of Statistics data (and a response rate of 89.7 per cent of 6,079 mining, manufacturing, and service establishments surveyed), the department found that training in industry was on the rise, but that only a third of the firms had organized training activity. The larger firms did the most training, the smaller firms the least. Still, only 7.2 per cent of the establishments employing skilled tradesmen did any training at that level, and only 3.6 per cent of skilled workers had access to training.

"Most employers may not need fully-trained craftsmen but instead workers efficient in a given set of skills," the report noted. "Private training in [establishments surveyed] is probably more for introducing employees to new work situations or production techniques

through short familiarization courses than for skill development."
Companies *with* formal training programs focused them to a disproportionate extent on employees involved in overseeing other employees. "Private training activity is largely conducted at the supervisory
and managerial level," the study concluded.[22]

When the government scrapped the TVTA and launched new
"manpower" programs in 1966-67 it moved firmly to a market
approach that would last for the next thirty years. It replaced the
TVTA's shared-cost federalism with a "purchasing federalism" program, under which Ottawa would stop handing money over to
provinces (which, like Ontario, could then use it strictly according to
their own priorities). This eventually turned into a seat-purchase
approach under which the feds provided funds according to the
number of places offered in the provincial education and training
programs of the new community colleges.

Tom Kent became deputy minister of the new Department of
Manpower and Immigration, and Canada's new manpower policy was
designed to achieve two main goals on the supply side of the labour
market: provide adults with training so that they could cope with the
technological conditions of the future and provide employers with
services that would "meet their needs for the right kind of manpower,
with the necessary skills at the right time and in the right place."[23]
The assumption, of course, was that the two goals were compatible.

There was also a recognition that the state had a far greater role to
play in assisting adults who had been let down by the educational and
training systems: allowances paid to adults in New Brunswick's retraining programs were as little as ten dollars a week. Under the federal reforms administered by Kent's rationalized department—which
was busy setting up 350 Canada Manpower centres to provide one-
stop unemployment insurance and job-finding services—trainees
also began receiving benefits competitive with unemployment insurance. Training ceased to disqualify people from unemployment
insurance.

By the late 1960s the groundwork was in place for a modern system of labour-market policies that focused in part on education and

training. Federal spending on training rose sixfold, from $50 million in 1966 to $300 million in 1971. Meanwhile unemployment remained below 5 per cent from 1964 to 1969 as the economy, spurred by the resource and automobile industries, continued to generate jobs. Welfare state industries such as health, education, and welfare continued to expand, offering up new job chances for women. Social workers, nurses, and teachers were hired in their thousands, most of them trained in publicly supported institutions—the universities and colleges that were growing like Topsy.

Some training dollars did find their way into the private sector, where companies continued to orient their activities towards their own particular needs as opposed to those of their workers. Spending money on this sort of training was—and remains—popular with government. Governments, eager to please their business constituencies, like to seem practical; businesses like to receive subsidies. Then there's the price factor. Training-in-industry is a good deal: "The much lower cost of apprenticeship training, or on-the-job training generally, as compared with institutional programs, has always been one of its most compelling features."[24]

Some twenty years later a Macdonald Commission study referred to the Pearson years as "the most active period in Canada's history for the development of social policies."[25] The Canada Assistance Act even had as its stated objective "the prevention and removal of the causes of poverty."

But despite economic growth and new social programs, poverty remained a stubborn fact of life. In 1968 the Economic Council revealed that an astonishing 27 per cent of Canadians lived in poverty. The Senate's Special Committee on Poverty, chaired by David Croll, issued a landmark study showing that poverty was not confined to people on welfare and unemployment insurance; but, rather, millions of "active labour market participants," people who worked at low-end jobs and/or worked sporadically, were also poor.

"Unless we act now, nationally," the Croll report warned, five million Canadians "will continue to find life a bleak, bitter and never-ending struggle for survival."[26]

The Croll report also pointed out that the poor, whether working or not, were also the people with the lowest educational levels. They were—and are—forced into low-wage jobs and vulnerable to unemployment. The Committee argued that the "failure" of the educational system to meet their needs was a waste of a precious resource. A real *human resources* approach would not be narrowly economic, but would instead emphasize the broader concept of citizenship and the "inherent value of a literate, educated and trained population" as well as the "social costs of under-development and under-employment of the nation's human resources." The committee rejected what it called the "narrower" approach to skill training "concerned with the development of human resources only in the economic sense—to increase productivity, and earning and spending power."[27]

The new Department of Manpower and Immigration (later Employment and Immigration and, by the 1990s, part of Human Resources Development), introduced a more comprehensive "manpower" approach. Under the Canada Manpower Training Program (CMTP) spending rose by 176 per cent in the three years ending in 1971-72, hitting $290 million in the last year alone. Much of this money went to the literacy and basic skills training emphasized by the Croll report.

In the early phase of the CMTP, counsellors channelled a large number of the poor people who were passing through the doors of the new Canada Manpower centres into general upgrading courses designed to help them with reading, writing, and arithmetic. As one evaluation of CMTP delicately put it, this approach reflected a "pressing need" for "an improved knowledge of language and other communicative skills, as well as a better grasp of elementary arithmetic and science." Nearly two out of every three authorizations for federal training between 1968 and 1970 went to such basic training or language training for immigrants.[28] This *equity* orientation, with its focus on disadvantaged groups, reflected at least a partial commitment to using training money to address the issue of social inequality.

This approach can be contrasted to an *efficiency* model of training. Directing training towards the development of specific skills tends to serve the goal of economic growth, often by tailoring the skills of individual workers to the express needs of employers. According to an analysis written in the early 1970s, the programs could indeed partially serve the goal of equity in helping to create "a more even distribution of income." But: "The persons selected for training will be those whose training most benefits the economy; such persons are not necessarily those in the lowest income groups. Instead, they are individuals whose training costs will be lowest in relation to the income gains that training enables them to realize."[29]

This "equity/economy" balance showed signs of tilting towards equity and basic skill training in the late 1960s. The skills emphasized were not tied to a single firm, but were portable. They would help particular low-income workers improve their general prospects. Fully half the recipients of basic training had less than two years of high school (though only a quarter were women).[30] At the same time Labour Minister Bryce Mackasey had fought off business opposition (and many cabinet colleagues) by pushing through changes to the unemployment insurance system that extended coverage, eased qualification rules, and boosted benefits. Mackasey was also a supporter of using training funds to help "the disadvantaged."[31] For those concerned with equity it seemed like a high point in Canadian labour-market policy. From the view of employers, it was like sinking into the deep mud of government interference.

However, while training and basic education were at last being conceived in slightly broader social terms and not strictly as means to ensure that companies had what they needed when they needed it from the supply side of the labour market, basic policy approaches never strayed far from the ideal of tailoring training to the needs of employers.

Although it pointed out the scandalously high incidence of poverty, the Economic Council was convinced that the way to clear up the problem was through economic growth that would trickle down to the unemployed and the working poor. The Council's influence was significant; its urgings had prompted the government to embark on an active labour-market policy in the first place. Noting that federal spending on training had increased sixfold between 1966 and 1971, the Council pointed out that "experts" were in general agreement that training in industry was preferable to institutionally based learning.[32]

The message did not fall on deaf ears. By 1971 the winds of monetarism and restraint were picking up, and Ottawa saw the costs of training in public institutions rising. Late that year, the short-lived caution about training in industry was relaxed, and the government began paying 75 per cent of the wages of unemployed people hired as trainees by private companies. Within a few months Manpower and Immigration was reporting that "the response from industry has exceeded all expectations," and the department increased funding for the program by 150 per cent.[33]

By that time some business voices were already sounding the alarm about the high cost of public education. As early as 1968 the Bank of Montreal, warning that education accounted for 40 per cent of Ontario's total budget, was urging a more spartan approach to spending. There were those in the bureaucracy who agreed that training was best left to the private sector. One internal report marked "restricted/for official use only" referred to a basic policy in the manpower department: "The continuing responsibility for training *his* workers must remain with the employer. Any attempt to turn this responsibility over to government is to be discouraged. ... Training should be conducted wherever it can be done most effectively and at least cost."[34]

Laudable as such sentiments may seem, the fact is that Canadian firms have had a dismal record in helping to develop the skills of their own workers. A comprehensive academic study of occupational training that appeared in 1973, in time to assess the first series of big government expenditures, offered the most charitable rationale for the business neglect of training. Stephan Dupré and his University of

Toronto colleagues indicated that training in industry had "operational problems" that were probably "endemic to the structure of the Canadian economy." The study concluded, "The small size and branch-plant status of many Canadian firms may constrain their capacity to mount efficient training programs."[35]

Other factors included the profoundly laissez-faire attitude of employers that had historically poached skilled workers from their competitors and relied on workers to train each other themselves, and relied on the state to tailor immigration policies to the business needs for skill. The government, for its part, had put in place an active labour-market policy concentrating on the supply side. The goal was to meet the demand for workers in both the emerging service industries and in manufacturing industries that were introducing technological changes requiring new skills.[36] At the same time government was displaying an ambivalent and tentative commitment to labour market spending that would concentrate on equity, preferring policies designed to stimulate the goal of economic growth.

By the mid-1970s Ottawa had begun to abandon Keynesianism in the face of a puzzling combination of inflation and unemployment. Both were on the rise, along with public debt. Reform-minded Liberals like Tom Kent (who had been at the centre of the Keynesian action) turned their backs on the Trudeau government in Ottawa, where conservative monetarist policies were becoming fashionable.

"The first requirement of social justice is that there should be enough jobs," Kent told Trudeau in a letter he prepared for his boss, manpower minister Jean Marchand. "That is a basic economic issue, but it is also the primary social issue." Kent realized that a major shift to the right had begun with the abandonment of the make-the-rich-pay equity proposals of the Royal Commission on Taxation, which reported in 1966. He felt that the monetarist policies that fought inflation and let unemployment rise would "turn out not to be a bottle of medicine but an alcoholic binge."

Against this background counselling and job training were futile. "There is little point in training people when people who already have the relevant skills cannot get work," Kent recalled at the end of the 1980s—when unemployment had reached a yearly average of 9.5 per cent, with worse yet to come.[37]

Despite this expressed futility emanating from many quarters, the emphasis on training as a resolution to unemployment and the shifting tide of work remained. "Human resource strategy is at the centre of our future competitive success," declared trade union intellectual and NDP civil servant Peter Warrian.[38] Apparently everyone agreed that there was a need for more training, whether workers were called workers or human resources.

Studies and reports and committees showed that Canada was an international training laggard. The tone was set by a high-profile document called "Adjusting to Win" produced by Ottawa's blue-ribbon post-free-trade committee led by Jean de Grandpré, then head of Canada's largest corporation, BCE. The de Grandpré report blamed Canadian business for not doing enough. "The private sector does not have a training mentality," it observed, raising the old issue of poaching. "If everyone were involved in such an effort, employers who do training would not see their staff raided by employers who do not invest in the skills of their workforce. . . . Training is not a residual activity. It must become part of work."[39]

In the late 1980s it was revealed that while the average Canadian worker received 6.7 hours of formal training, the figure in Japan was 200 hours. Canadian firms spent less than a quarter of the amount laid out by their German competitors on training.[40] Overall, firms probably spend more on computer paper than on training. So pervasive is the recognition that Canadian business has failed on the training front that the president of Xerox-Canada has said that the average Canadian firm spends $98 annually to teach new skills to its employees—"less than the cost of dinner at a moderately good restaurant."[41] While this says much about the dining habits of corporate grandees, it also shows that Canadian business is still suffering from a laissez-faire hangover brought on by decades of importing and poaching its workforce and of allowing the state and individual workers to provide the necessary skills. But, as the de Grandpré report concluded, Canadian business was not about to change its ways "simply because it is exhorted to do so."

Who should pay for training? Labour has traditionally countered management's traditionally laissez-faire approach with the idea of a training levy, a payroll tax that would be fully refundable to employers who actually undertook active training programs. Although the private sector subscribes to the training-equals-competitiveness gospel, it has regarded as subversive the idea of a training tax, even one that is refundable to those who back their ostensible commitment to training with actual practice. This would, after all, constitute meddling with the operation of the free market.

Business has had its way. In Ontario, after a complex political process during the early 1990s in which the idea of a training tax was often raised, the (NDP) government of the day failed to introduce an employer training levy, even though the idea of a training levy had been endorsed by the Mulroney government's de Grandpré report. Nor did Ontario come forth with legislation that would have required employers to train workers.[42] Instead, it decided to foot the bill for a new provincial Training and Adjustment Board run jointly by business and labour with some participation from "equity" groups—women, visible minorities, the disabled.

This kind of neocorporatism appears to be the wave of the future in Canada. Individuals, employers, and the state will all share the costs of training, with the exact division being the subject of continuing political dispute. Business can be expected to support a voucher-style system based on a market model in which individuals choose their training from both the public-training and the growing private-training sectors. Labour will support training in public institutions.

In the end, though, who pays for training is a much less significant matter than who gets trained—and what they learn.

6

PRIVATE TROUBLES IN THE BORDER CITY

The doctrine that if the horse is fed amply with oats, some will
pass through to the road for the sparrows.
> – Trickle-down economics, as defined
> by John Kenneth Galbraith, 1992

MIDSUMMER evening in Windsor, and the sun sets late
over the river and the devastation of Detroit's near west
side. The annual cross-border Freedom Festival is in full
swing. The downtown streetlights are adorned with bunting emblazoned with a complex logo that combines the stars and stripes and
the maple leaf. This year the Freedom Festival coincides with a convention of four thousand Baptists being held in Detroit, and the
spillover has to be accommodated in hotels on the Canadian side.
School buses from churches across the United States are parked in
downtown hotel lots.

The Americans who come to Windsor for the Freedom Festival
don't penetrate the residential heart of the city, instead sticking
close to the downtown tourist bridgehead with its bingo parlours,

restaurants, and safe streets. Down along the riverside a midway is operating in the narrow strip of shoreline parkland. A powerful launch roars up to pick up some Coast Guard men who have been partying in Windsor. Cheetah's strip bar is close at hand.

A big attraction at the Freedom Festival midway is the Casino Tent. Blackjack predominates, and although the players seem rather joyless, a volunteer dealer at a two-dollar table predicts that when the real casino opens up you'll be able to make eighteen dollars an hour and they'll need two thousand dealers. Soon the Festival organizers will be competing with the glitz of the Ontario government's *real* casino just downstream in a converted brewery warehouse.

Over at the ten-dollar tables the dealers have been drawn from the ranks of the first intake of students at the new "gaming" course offered by St. Clair College. One of them is a hairstylist and Windsor native named Daryl who runs a salon near Ouellette and Wyandotte, a main downtown intersection. Like the other dealers, Daryl is wearing black pants and a neat white shirt with a small bronze ID tag that reads "St. Clair 0116." He points to the 01, explaining that it shows he's part of the first class at the community college. Daryl feels that he has gone about as far as he can in the salon business, even though it sometimes seems that the only secure trades might be barbering and undertaking. Daryl wants to try something new and is hedging his personal bets in the job sweepstakes on another skill of the future.

"Once you learn the tricks of the trade, it's a doorway to anywhere," he says. Indeed, the industry seems to hold out abundant hopes for a growing number of cities across the continent. "It's a wonderful thing for Windsor." Daryl has been a keen learner. He explains that the key to dealing blackjack lies in the ritual and "etiquette" of the game rather than in just knowing "when to hold them and when to fold them."

"It's like a production line at a factory," he says off-handedly, flipping out a card. "What keeps it different is the players."

Further east, between the GM Trim Plant and the river, Margaret Peever sits at home. She's been to St. Clair College, too. "People are *paying* to learn to deal *cards*," she says. "People are talking about getting jobs cleaning."

Peever is scared of what the future holds for her. Her job racking and packing auto parts disappeared many long months ago. Ever since then she's been looking for a job, and taking courses. She lives in a suburban area that the developer called the "Villages of Riverside" when it packed dozens of modest structures onto tiny lots using a euphemistically named "zero lot line" concept. The locals call the settlement "Apache Village" because the tall, stockade-like fences dividing up the properties resemble the forts in which the U.S. cavalry took refuge from the Indians in hundreds of old western movies.

Margaret Peever moved to Windsor in 1965 after dropping out of high school in North Bay, Ontario, where her father and two brothers were bricklayers. She was soon married to an autoworker, raising four children and moving in and out of the various jobs that women have traditionally held in Windsor—power sewing at GM Trim and working as a cashier and sales clerk at Woolco. When her twins were old enough for day care, she landed a job at FCM, a small factory that the giant Gulf & Western conglomerate had just organized as a branch of its Wickes Windsor Bumper. FCM made plastic grills and other parts. Peever chromed plastic parts and did spray painting. Even though it was unusual for women to be employed at this kind of plant, the work was steady and the pay decent enough—though lower than assembly work on the line at Chrysler where her husband worked.

By 1980, Windsor's auto sector was reeling from a massive downturn that combined double-digit inflation and an unemployment rate that hit 12 per cent. Plants were closing all over the region, and people were losing their houses and heading out to Alberta and B.C. and the promise of work in the resource industries. In May Gulf & Western announced that Peever and the fifty-seven other workers at FCM would soon be out of work. The company also threatened to close the main Windsor Bumper plant unless the workers agreed to a concession package including a reduction in cost-of-living increases.

Faced with high inflation, the workers were less than happy with this proposal and responded with a spontaneous occupation of the bumper plant. This was part of a spate of factory occupations in the early 1980s, with nothing-to-lose workers using the old sit-down tactics of days that had seemingly gone past. When hastily scribbled placards ("We'll go to jail for our jobs") appeared at the plant windows and supporters smuggled in supplies, the occupation seized the imagination of a city on the defensive. It also surprised both management and the United Auto Workers Union, which started to negotiate. Within a week the occupation was over. The resulting deal brought some gains for the workers, kept the plant open, and—most importantly for Margaret Peever—offered preferential hiring at Windsor Bumper for laid-off FCM workers.

It should have been a happy time for Peever. But during this period her marriage broke up. At the same time it became clear that the deal with Gulf & Western promising FCM people work at Windsor Bumper did not include the handful of women who had been working at the plastic plant. Peever reluctantly decided that she would go back to school, beginning a career in retraining that would continue for over a decade.

She also hauled Gulf & Western (a company that had gained notoriety for turning the Dominican Republic into a corporate client state) before the Ontario Human Rights Commission. Peever and nine other women told the Commission that the company simply would not accept their applications. "They figured women couldn't do the work," she says. Hoisting metal bumpers around was admittedly heavier work than dealing with the plastic parts at FCM. But, she says, "All we asked for was a chance."

After a three-year legal struggle Peever got her chance. The women forced a settlement out of Gulf & Western that gave them a choice between a cash buyout and jobs at Windsor Bumper. The conglomerate also agreed that it would not force the three women who took the job option to work racking heavy bumpers on a continuous basis and that it would embark on an affirmative action plan to bring more women into the good jobs at the plant.

By the time she finally went to work racking bumpers, Peever had completed studies for her high-school diploma and learned how to design flyers and posters as part of a job-training course sponsored by the YMCA. Even though she now had a job, she had developed a thirst for learning that would not go away. She began to take training courses sponsored by the autoworkers union, courses that taught her to deal with health and safety issues, act as a counsellor to other workers in their dealings with the unemployment insurance bureaucracy, make a speech, and act as an effective union steward: leadership skills, social skills, the skills needed to participate on the job and as citizen-at-large.

"When I finally got into Windsor Bumper, I thought that was my future," she says. "That's why I took union courses. I thought, 'This is it. I'm secure, making good money.' I bought a house."

But in 1990 Windsor Bumper closed along with dozens of other plants in Windsor's auto sector. This time it was something more than one of the periodic downturns that have always hit the city. Analysts put it down as the permanent result of a sour mixture of free trade, recession, automation, and corporate restructuring. This time many jobs simply were not coming back.

"Now I'm looking at losing my house and everything," Peever says. At forty-five years of age, living alone (her children have headed off into the world of work and education), she has fears about what the future holds in store for her. She is afraid of losing her home. Still, she describes herself not as unemployed, not even between jobs, but "between courses." She has a stack of certificates and diplomas sitting neatly on her kitchen table. One of the certificates is for a "Women In Metal Machining Course" at St. Clair College.

Like other working-class people of the 1990s, Peever has been persuaded by various counsellors to put together a resumé. A single page summarizing her twenty-plus years of working in a half-dozen jobs—polishing, sewing, making change, spray painting, inspecting, racking, assembling—the resumé says she is a "flexible employee," willing to work shifts, "self-motivated." Again, it is as if workers were empty bottles constantly in need of refilling with skills, the hard skills for specific jobs and the soft skills for the right attitude. Sitting on top of her neat stack of qualifications is the black tasselled mortarboard hat that Peever wore when she received her academic high-school diploma.

Discussing her plans for the future, she says that this time she intends to stay away from factory work. It is not so much the physical strain on a middle-aged body—a tall, strong woman, she was able to cope with heavy work—but there were other drawbacks to life in the plant. She picks up a file from a sexual-harassment course and extracts a cartoon she had pulled off the wall near her work station.

"Working with men takes a lot of stamina," she says. The crumpled cartoon is a full page in colour from a "men's magazine" of the *Hustler* variety. It shows a naked woman reclining alone on a bed, legs spread. Unlike the young models who undoubtedly fill the other pages of the publication, the woman in the cartoon is middle-aged, with thick thighs and hips. There is a vibrator running away from the bed, vomiting as it goes. "Nice, eh?" Peever mutters.

Peever says she enroled in a medical lab-technician course at St. Clair College in the hopes of escaping factory work. She has heard that even though Canadian hospitals are shutting down entire wings and cutting staff, there is a chance of finding work in the United States if you have the right qualifications. She doesn't want to move from Windsor, where her friends and family live, but, she says, she would move to Timbuktu to feed herself. She is willing to give retraining another try to qualify as a medical technician. Still, competition for spaces in the course was intense. It is not like before, when she first went back to St. Clair. Then, she says, "They would pretty well accept you with any background as long you had the initiative to learn."

Hoping that it would be useful in her lab-technician studies, she took a biology course offered by the local school board. But soon after the community college course began she realized that many of her fellow students already had university science degrees. She found the struggle to keep up with the academic pace unnerving, the workload tremendous. She was nagged by the anxiety of being out of her depth. A few months into the classes she sprained an ankle getting off a bus on the way home from school. The few days of school she missed were enough to persuade her that she would not make it as a lab technician.

She lights a cigarette and tosses her head back with laughter, seemingly at ease. Clearly, she has the initiative to learn. She says that Windsor has good networks made up of women who share her situation and talk regularly, go for long walks, and offer each other support. She likes biographies, so she spends time reading as well as crocheting and looking after her cats.

But the "reality" is starting to get to her, she says. She is worried that the reality might include moving to the States. "Windsor is my roots now—this is my home town. It's frightening because I'm not young. I'm not married. I don't have anyone to take along with me. I'm completely on my own. I'm tired and I really don't want to start again. But in order to survive I've gotta keep it in my head that I might have to. And that's a hard thing to do."

For Margaret Peever, the brave new world of work is a place that brings up a complex jumble of emotions. She is, she says, sad, angry, hostile, and resentful, all at once. "You have to have a job to feel good about yourself. Not everybody can be on welfare. Where's the money coming from? They talk about lowering the deficit, well who's going to lower it if we're all out of work?" She says she has done everything "they asked" her to do. "I got retrained. What do they want me to do now?"

According to conventional social policy wisdom, someone in this situation has to reconcile themselves to what the job market has to offer—which means a lot of part-time and temporary work. For those who once held jobs that came complete with pension and dental coverage, it can be bewildering. "There's no job security. There are fourteen people behind you waiting to take your place as soon as you make a mistake, or if you don't do exactly what you're told. You do everything they say and smile and be grateful for them asking you to do it."

When Margaret Peever began working in the factory after her youngest children passed the toddler stage, it was already becoming clear that precarious and part-time work was the order of the day for working women, particularly those in the service sector.

In 1980 the minister of employment was Lloyd Axworthy, who would hold the same post fourteen years later in the government of Jean Chrétien. He told a meeting of the Organization for Economic Co-operation and Development (OECD) that although Ottawa was making efforts to get women into non-traditional jobs in manufacturing industries, the real challenge was to improve the conditions of part-time work, because it was characterized by "low pay, unsatisfactory working conditions and lack of access to social security and related benefits." Axworthy insisted that Canada had to make sure that "part time workers do not continue to be viewed as an available and reserve source of labour that has less need of benefits."[1]

Women had been stereotyped as housewives in the modern era. So it was assumed that they could work like housewives in the formal labour market; that is, for low pay. It was also assumed that most female workers had access to support outside the labour market.

In the past decade or so, the flexible, just-in-time worker has become the management ideal. Service-sector firms with low ratios of capital to labour find the key to increased productivity is to squeeze their workforces into cheap, part-time, as-needed positions.* Thus the apparent paradox of politicians being able to point to solid job creation while unemployment remains high. More people work at "jobs, jobs, jobs" that give them a form of subemployment.

Not all observers see the future in such terms. In an article entitled "Working-Class Heroes," a national business magazine served

---

* There are parallels as well between the jobs of today's idealized flexiworker and the work of the housewife. Both are low-paid (or no-paid in the case of the housewife). Both are expected to be flexible, able to respond to changing needs by undertaking whole new sets of tasks. In the case of the housewife, these include the tasks of nursing, chauffeuring, counselling, cleaning, cooking—the list is almost limitless.[2]

up a typical description of "a win-win" situation in which workers and employers get together to "kick-start a struggling economic recovery." The "squeeze" in question was not the shrinking middle, but the shortage of particular skills faced by some corporations. The upbeat message was that although unemployment is high, employers cannot find skilled workers because of a "cultural bias" among Canadians who think manual workers "are stupid."

The answer? Ignore those old-fashioned attitudes and get with the flow: get trained as an industrial machinist or a software engineer. We are presented with a forecast that unskilled jobs will decline to a mere 15 per cent of the total by the year 2000. The rest of the labour force will apparently be employed as engineers and machinists. The answer, somehow, is for individual Canadians to gear their training strategies to take advantage of the "glaring paradox" of high joblessness and employers "desperate" for skilled workers.[3]

Talk of the high unemployment/skill shortage paradox is not new. In 1978, when unemployment hit a post-Depression high of 8.3 per cent, Windsor was suffering from a major shortage of skilled trades and company owners were crying the blues. Tool-shop owner George Shaffer claimed that if Windsor, the heart of the skilled labour force in Canada, was in trouble, there must have been a shortage of skill right across the land.[4] A business consulting firm pointed to an expenditure of $2.5 billion in public training over five years, high unemployment, skill shortages, and the need for "a massive overhaul" of job training. The consultants said it was "disgraceful" that people could not get into skilled trades. A business columnist concluded that it was "tempting to call this the scandal of the seventies."[5]

At the same time, studies from the files of the Ontario education ministry and Ottawa's employment and immigration department indicated that, in the Windsor area, less than a quarter of all skill requirements were being met by in-plant training, a fact that revealed the "weakness" of private sector training.[6] Windsor's Chamber of Commerce set up an emergency task force on training and the city council, with one eye on the historical record, sent a letter to Ottawa saying, "The only alternative to improving the qualifications of local citizens is to encourage the immigration of skilled persons from foreign countries."7

But Canada's traditional sources of skilled workers now either had powerhouse economies of their own (Germany) or were facing their own problems producing skilled trades as their industrial strength waned (Britain).8 What's more, the managerial approach of the Pearson Liberals, who had operated largely on the basis of labour-market principles, had given way under pressure from human rights and minority groups to a 1978 Immigration Act that included more liberal and nuanced policies emphasizing non-discrimination, the acceptance of refugees, and a general humanitarianism. Not to be excluded, of course, were pragmatic economic considerations related to the fact that skilled workers were now only to be found in places like Asia, from whence newcomers had historically been restricted by nativist immigration policies.9 The new act attracted a good deal of public attention.

That same year, 1978, also saw Ottawa make another major policy change. That summer Prime Minister Trudeau returned from a summit in Bonn that had accepted the OECD's influential McCracken Report entitled—with extreme irony given the benefit of hindsight—*Towards Full Employment and Price Stability*. In line with the report, Canada had agreed to take more monetarist medicine, to cut back on public spending and reorient it to the private sector in the hopes that business would create jobs and train people for them. Public labour-market policy was, in the words of one review of the period, "under attack."10

Ottawa had begun a trend that was to continue for the next fifteen years. Finance Minister Jean Chrétien immediately announced cuts in benefits paid to the unemployed under UI and increases in eligibility requirements and minimum earnings needed to qualify for benefits. Employment Minister Bud Cullen told the provinces in September that training allowances to the unemployed would be "drastically cut" and that the new idea was to "shift the emphasis to helping industry train and employ Canadians."11 In spite of industry's dismal training record, the government enthusiastically embraced private-sector training, announcing a new Critical Trade Skills Training program aimed at subsidizing employer costs in skilled manufacturing trades.

The supply-demand approach to the job market was neatly summarized in a paper presented by the Canada Employment and Immigration Commission (CEIC) to a First Ministers' Conference that November: "On the demand side, emphasis should be shifted to private sector employment development. . . . On the supply side continuing attempts should be made to increase incentives to work and remove barriers to employment which could arise through a lack of appropriate skills."[12]

An indication that the long-standing neglect of training by both government and industry had not been addressed came in a 1979 cabinet submission written by T. Philip Adams, Ontario's deputy minister of colleges and universities. Adams told the provincial cabinet that employers had for many years been "delighted" with the "no-cost, no-risk solution" of importing skill. "The Government's action: in the face of indifference on the part of industrial employers and no interest on the part of the public nothing much seemed warranted and, until recently, very little was achieved."[13]

Historically, the discovery of the simultaneous occurrence of high unemployment and skill shortages was usually followed by discussions of male-dominated, skilled manufacturing jobs that were almost invariably full-time jobs. But in the late 1970s and into the 1980s, part-time work in services, generally women's and young people's work, was the fastest-growing part of the labour market. Between 1975 and 1986, according to Statistics Canada, involuntary part-time employment (or, more properly, subemployment) grew by an astonishing 375 per cent, while voluntary part-time work rose by 41 per cent. Lagging well behind was the rise in full-time jobs: 15 per cent.[14] Plans for workplace training were obviously neglecting the needs of the new labour force. Part-time work even continued its growth during the depression of 1981-83, a period that witnessed an actual decline in full-time workers.

It was at this time, with a million and a half Canadians officially unemployed (a rate of 11.9 per cent), that two important reports dealing with the choices for labour-market policy came out of Ottawa. They had markedly different perspectives. A parliamentary task force study (*Work for Tomorrow*) under veteran Liberal Warren Allmand supported the notion of full employment ("that is, a job for every person willing and able to work") and pointed out that unemployment could be attributed largely to insufficient job growth.

"In the absence of a government commitment to full employment," the Allmand report concluded, "it is impossible to plan and carry out job creation and training programs which will be of any benefit."[15]

The Trudeau Liberals and the Mulroney Conservatives after them ignored this obvious fact in favour of the approach recommended by the second study, the Dodge report, so named for its author, a senior civil servant. While acknowledging that structural unemployment (joblessness unrelated to the swings in the business cycle) had risen, this 1981 report placed the blame on the recent more generous unemployment benefits, such as the Mackasey reforms of 1971, as well as on demographic shifts and on the fact that more women were entering the labour force. It maintained that unemployed people no longer experienced the "hardship" they had in the past, presumably a reference to the Great Depression. The Dodge report was so out of touch with reality that it actually predicted major labour shortages by the mid-1980s.[16]

This approach simply put wind into the sails of a ship of state already drifting steadily, under the Liberals, in the direction of laissez-faire. By the beginning of the last Trudeau government, Thatcher was already in office, Reagan was on his way to the White House, and right-wing thinking was sweeping the last vestiges of Keynesianism out of the corridors of power, particularly in places like the OECD. Coming back in was the liberalism of C.D. Howe, the man who dominated Ottawa for so many years and for whom the notion of anything but business as the agent of change was anathema. Howe had resisted any explicit policy goal of full employment. Indeed, against this background it seemed that by the time the Mulroney

Tories took over in Ottawa in 1984 the brief interlude of the partially erected welfare state had been a historical hiccup.

The Tories continued the Liberal cuts to unemployment insurance and re-emphasized training as the key to the future. By the end of the decade they were appropriating hundreds of millions of dollars from the social safety net side of the UI fund and diverting them to training and the Unemployment Insurance Developmental Uses side. In 1990, 20 per cent of all training money came from the UIDU fund; by 1992 the figure was 60 per cent.[17] In this way Ottawa could claim it was spending more on training, but in fact it was cutting back on both the UI fund *and* eliminating its own contributions to UI as part of its battle against the deficit. All of this was dressed up in programs with names like "Adjusting to Win" and "Success in the Works," which paid no attention at all to the dilemma of polarized growth. As Rianne Mahon of Carleton University's policy school pointed out at the time, the policies presented a distinct danger that a certain group of "core workers" would get quality training while the "new poor" would be taught to accept their lot in the unstable, low-wage jobs of the new economy.[18]

From the time the Conservatives took office in 1984, an open-for-business vocabulary dominated discussions of how to get Canadians back to work. In the era of deregulation mania, policy would be driven by the notion of "letting the market decide" just about everything. It would be up to the private sector to make the investment decisions that would spur economic growth and hence boost the demand for jobs.

The state concentrated on the supply side of the labour market, pushing a renewed training agenda. In spite of massive evidence to the contrary, the government kept insisting that unemployment had nothing to do with basic changes such as the microchip revolution and the replacement of people with machines, or the switch to a service economy that was fracturing work along good jobs/bad jobs lines. Nor could it have been related to corporate strategies of abandoning any vestige of commitment to job stability in favour of the trend to a just-in-time workforce. Likewise, the relocation of jobs to cheap labour havens was not a cause of structural unemployment.

"Much of the increase in unemployment," claimed Health and Welfare Minister Benoît Bouchard at a Paris meeting on employment policy, "is structural in the sense of mismatches between the existing skill-set of workers and the skill requirements of jobs." Bouchard was introducing "new orientations" on social policy that had actually been in place since the late 1970s. The principle guiding Canadian employment strategy, he said, must be "individual responsibility."[19]

———————

Two months after Brian Mulroney's Quebec lieutenant flew to Paris to underline his government's commitment to stout-hearted self-sufficiency, the staff at Windsor's Unemployed Help Centre (UHC) arrived at work to hear the news of the prime minister's resignation. They quickly got busy and had soon stuck up a large sign in the front window of the office: "GOODBYE LYIN' BRIAN, COLLECT YOUR UI WHILE YOU STILL CAN."

The front of the Help Centre stares out at the office of the busy Canada Employment Centre and, beyond that, the old Champion Spark Plug plant, darkened now, its equipment shipped out to a Champion operation in the States. The Centre is a huge, labour-backed operation with a staff of thirty-four who offer everything from basic literacy to computer courses. It provides counselling for the unemployed, boosting self-esteem and teaching social skills. The UHC even co-ordinates the central depot that supplies Windsor's growing network of food banks. It is housed in a non-descript, low-rise structure not far from the massive Ford engine plant on the city's east side. The building formerly held both the Paradise Wedding Chapel and a failed warehouse operation. Jobless people, from teenagers to laid-off workers in their fifties, file into the back half of the building that has been turned into classroom space.

I sit in on one of the "Basic Skills" courses at the Help Centre. The Anglo-Canadians outnumber recent immigrants, but just. A serious young woman named Asha frowns into her workbook. Back in Guyana her parents invested what money they had for the education of her brothers. She started working at the age of ten. Now she's taking a crack at a business letter. Asha is quiet-spoken but, as it soon becomes clear, she is the ideal learner. In her nine years in Canada she has taught herself to type using a manual typewriter, and she is now volunteering as a home-care worker, visiting lonely old people and helping with housework in the hopes of beefing up her resumé. Asha admits she still has trouble with math and writing.

"There's no sense rushing out now, without the proper schooling," she says. She tried to get a job as a waitress but was told she had to have her Grade 12. "They always want skill."

A chunky young man wearing a Detroit Red Wings sweater talks with a mixture of hope and nervous bravado about the situation in northern Ontario, where, rumour has it, the lumber mills have picked up and the welfare officials "don't give you no jack around."

The learners—twelve of them—sit at tables and veteran literacy instructor Judy Fortin circulates among them, helping with spelling and syntax. There's a tile floor, grey cinder block walls, no windows. Fortin knows that 28.4 per cent of Canadians are functionally illiterate—something the Allmand report emphasized in its plea for an equity approach to training. But, as with teachers in classrooms everywhere, she doesn't enjoy the undivided attention of all the students. They are easily distracted by the chance to swap yarns about the world of working—or not working.

"Now if you don't have a resumé and job skills they won't look at you," explains Joan, a women from Edmonton who started working as a dishwasher at fourteen. She is thirty and has Grade 9. This morning she dropped her resumé off at Gold's Gym, where they had an opening for a part-time juice bar attendant. "They're going to hire a university student over me even though I have experience," she says between bouts of a cough that sounds like breaking glass.

"There's a lot of competition. That's the problem," says Dwayne, who undoubtedly has trouble identifying with the pronounced business-page virtues of the healthy new age of competition. He's had trouble besting his fellow job searchers.

Another student, Carlos, fled to Canada from El Salvador during that country's civil war. He has three kids and has just quit welfare because the authorities demanded that he sell his car. As a town dedicated to the automobile, Windsor's public-transit system rivals that of Los Angeles. You need a vehicle if you want to conduct any sort of active job search, let alone get a family of five around town. Carlos spent four years working full-time making EverFresh Juice for seven dollars an hour before quitting to take a nine-dollar job at an auto-parts plant. Unfortunately, the parts plant was one of the many that disappeared in 1991, and Carlos became one of the many who were tossed aside like used Kleenexes. When the job recovery was slow to arrive, he ended up applying for welfare. His wife supports the family with a nurses' aide job in a retirement home, and this clearly irritates his sensibilities about the male being the family breadwinner. "Nobody's gonna train you," he says bitterly. "They just want experience."

Unlike the dutiful Asha, Carlos has an apparent attitude problem. He is a prime candidate for the attitude adjustment emphasis that's so strong in worker training programs. He was one of the hundreds who lined up overnight in the rain hoping to get an application for a job at Chrysler and a chance at the grey monotony of the assembly-line. He was one of the lucky ones who got an interview before being rejected. It seems Carlos was too small for work in an industry that claims it is looking for lively minds and no longer hires "from the neck down."

"Carlos," says Joan, the Basic Skills co-ordinator at the UHC, "you have to realize that these are hard economic times. In the past if you could do the job and if you had no communication skills it was okay. Now it's very competitive. If someone can *choose* a Grade 12 they will. It's going to keep getting tough and competitive."

One of the other women pipes in to warn Carlos about the private job marts whose "employment services" aimed at recent immigrants consist mainly of selling photocopies of the want ads. "Why does the government let them do this?" Carlos asks. No one replies.

The street-side offices that house the Help Centre's staff are separated from the back half of the building where the programs take place. At break time the Basic Skills people file out to the lunchroom where they mingle with other trainees—people taking computer upgrading or printing courses, or who belong to the Job Club, which trains the unemployed about how to look for jobs. The lunchroom is much the same as in any workplace—neatly arranged but plain tables and chairs, microwave, fridge, vending machines, and so on. Pam Pons, the Centre's boss (or "ED"—Executive Director) has found out that the working-class people who use her operation tend to smoke, so she has made ample provision for smoking areas.

The lunchroom is where people gossip about their job searches, passing on rumours about places where jobs might or might not be coming up. On a bulletin board, beside a poster for an upcoming Wrestlemania card at the Windsor Arena, someone has tacked up a "Report on Business" article from a series on "The Jobless Recovery." The editors have chosen a special little logo for the series. It juxtaposes a PC terminal with a couple of men leaning idly against a wall, hands in pockets. Amidst the clichés about learning a living and the skills mismatch, the Windsor unemployed read that Statscan is predicting that non-standard work is at 30 per cent and rising, with traditional full-time jobs on the wane. They are told that the least promising careers are in manufacturing and that, according to former Economic Council of Canada staffer Gordon Betcherman, "There was a time when a good education and good training guaranteed you a good job. It's no longer a guarantee, but if you don't have it, you're guaranteed of getting a bad job."

That is, if you get any job at all. The clippings on the board also greet the trainees with "Welcome to the 10% Decade" and "Corporate productivity drive swells unemployment ranks: analysts believe many positions are gone for good." It is all very sobering, but the most stunning quote is from the premier of New Brunswick. "I have an absolutely dominating belief," says Frank McKenna, "that in this chicken-and-egg conundrum of whether you should have jobs or training first, the answer is that you need the training first. *If you have the training, the jobs will take care of themselves.*"[20]

This bizarre bit of logic might be amusing if it didn't come from the politician who seems to represent the vanguard of public policy thinking. Such dogmas are a variation on the old theme of trickle-down economics, the idea that growth is by itself sufficient, and all will benefit from it. A rising tide lifts all boats, and so on.

The growing ranks of the functional underclass come to Windsor's Unemployed Help Centre. Many have been members of the shrinking proletariat, people who had good, unionized jobs with good benefits in plants like Champion Spark Plug. Others, like Carlos and Joan, have never made it that far. Some have held office jobs. Most envy the workers at the giant Chrysler minivan plant, where the Canadian Auto Workers still flex their muscle. All have one thing in common—a pervasive sense of insecurity and even fear of the future. They come to the Centre in search of counselling, literacy, the training that they are told is vital to their survival. Many arrive in search of food.

Pat Kelly came to the Centre's Job Club sessions when his work as a sign painter disappeared. His organizational skills landed him what looks like one of the secure jobs for the future. Kelly co-ordinates the food bank. His job is to arrange bulk transport from Toronto's huge Daily Bread Food Bank when surplus vegetables, corn flakes, frozen food, or any of the other food bank staples become available. Small warehouses like the one Kelly runs dot the urban landscape; activists like Kelly regularly point out that food banks outnumber McDonald's outlets.

"It's a big job," says the thirty-something father of two, who arranges distribution to other Windsor-area food banks. "And it's getting scarier." Kelly points his thumb in the direction of Detroit, a murder city where the homeless haunt the abandoned streets after dark.

The Help Centre provides job counselling to everyone who receives a basket from the food bank. Kelly takes out a pad and begins to scribble down figures, telling me about a cycle he has seen repeated dozens of times. Kelly has been there. He first learned of the Help Centre and the welfare cycle when he lost a job in the recession of the early 1980s.

"Take some fellow," he says, "who's been working at Kelsey for twenty-three years." Kelsey-Hayes is a wheelmaker that closed and moved south after decades in Windsor. The worker, Kelly says, probably used to be grossing $40,000 a year. Now, on UI, his cheque could be about a $1,000 a month, and he might be spending about $650 a month for rent. He spends an entire year on UI getting a thousand a month, thinking, "This is the worst it can ever get." When his UI runs out he goes to social assistance. He's already spent his last UI cheque. When he says he doesn't have enough money for food, social assistance tells him to go see a food bank. Kelly says it all makes those seven-dollar-an-hour jobs look pretty good—that is how wages get rolled back.

Over on the training side of the Help Centre, Noella McWha has enroled in one of the thousands of computer training courses offered in pre-millennium Canada. She qualifies for the federally funded course because she is on UI. With twenty-five years' experience at shipping, secretarial, and other office work, she's trying to avoid the downward slide. Her last good job ($15 an hour) ended when the firm she worked for folded. Since then she has worked as a temp for a bank, a golf club, and a moving company.

Today McWha has just heard that out of eleven hundred applicants she was one of the lucky two hundred people to be interviewed for seventy-five temporary jobs cranking out parimutuel tickets at the Windsor Raceway. She has a husband on permanent disability. "You have to lower your standards," she observes matter of factly. She has a pragmatic view of training. "This course will make me more marketable."

The course has eighteen weeks of sessions. The atmosphere is different from what's going on at the austere Basic Skills course down the hall. There the learners sit on ergonomically sensible chairs in front of their terminals. The light is indirect. The floor is carpeted. The participants all have a basic familiarity with computers. Only one of five who wanted to get into the course made the grade. A keyboard speed test is part of the qualification ritual. Now McWha and the others are learning standard office programs—Lotus, WordPerfect, Bedford. There is one male student.

Noella McWha's best friend in the course is Mary Ann Lee, who has experience as an office manager at various Windsor-area trucking firms, doing payroll and invoicing work. Lee's working life has been interrupted by the arrival of her three children, aged ten, six, and two. Her husband, an estimator in the construction business, is unemployed. She hopes that the course at the Help Centre will broaden her computer skills; the trucking operations where she worked all had dedicated systems, so she could not transfer the skills she picked up there to other workplaces.

"It's scary out there," Lee says, thinking about the jobs "out there" and the interviews she will have to go through, giving "110 per cent." She conceives of herself in market terms, as a commodity. "This course should up my worth."

Both McWha and Lee are hopeful because, when they finish the course, they will know their way around a few standard office computer programs. But they also have an easy cynicism. For them, the union-sponsored Help Centre is just another in a long string of workplaces, with its own hierarchies and office politics. They even refer to it as "this company" and to the women (the Centre is a woman-run operation) who direct it as "the bosses." They feel that the UHC is playing a numbers game, trying to grind out as many trainees as possible, trying to be well-positioned for the next round of inevitable training grants.

The cynicism masks a sense of powerlessness, and an unmistakable anxiety. In their conversations with me McWha and Lee return again and again to the demand side of the job market. Although they are convinced that they must hone their own skills, the constant question lurks: "Training for what?"

This bedrock problem is compounded by an awareness of other difficulties faced by other women, women they have met in other offices and training courses. Lee and McWha talk freely about problems women have with alcoholic parents and abusive relationships. Lee remarks on the disadvantage that women have from the very start because they get paid less than men, and points out another hurdle. "We're a minority, white females," she says. "If I was a black female, I might have a better chance of landing a job. Equal opportunity employers—if you're a minority or have a second language you have a better chance. . . . I'm at the bottom of the list. The only person below me is a white male."

In the end Lee figures that the computer course is necessary not really to improve her skills, but "to keep up with the Joneses." Both women worry that if and when they do get work it won't pay much more than what they were earning ten years ago. They have paid their dues and have experience, and don't want to pay those dues again. The problem is, as Lee says, "Lots of people have the same skills we do."

———

Troubles, observed C. Wright Mills, are private matters. Issues, on the other hand, are public matters.

When Mills wrote this, in 1961, a particular kind of trouble, unemployment, had hit a postwar high of 7.1 per cent in Canada—a pinnacle it did not reach again for fifteen years. Then, from 1976 to the mid-1990s the unemployment rate has never dropped below 7 per cent. By 1994 the finance minister's statements were indicating that a rate of 8 per cent was a lofty goal. Unemployment in Canada has been consistently above that of other OECD countries.

Mills put it all into perspective: "When, in a city of 100,000, only one man is unemployed, that is his personal trouble, and for relief we properly look to the character of the man, his skills, and his immediate opportunities. But when in a nation of 50 million employees, 15 million are unemployed, that is an issue, and we may not hope to find its solution within the range of opportunities open to any one individual."[21]

Yet in Canada there is an overwhelming emphasis on individual responses: personal skills, hard skills, soft skills; training, retraining; adjustment, adjusting to win. According to the Mulroney Tories, the answer was "individual responsibility." According to their successors, it is "self-reliance."[22]

A 1991 Economic Council of Canada study of Canadian employment policy concluded that conservative thinking dominated Ottawa's policy from the early 1980s onwards. "Increasing individual initiative [has] been the predominant political and economic discourse in this area, with employment considered to be a market matter."[23] Unemployment becomes private trouble.

For more than a decade now, for instance, job clubs, pioneered by the Canadian Employment and Immigration Commission, have been commonplace in the Canadian experience of unemployment. Job clubs do training sessions designed to convince the unemployed that "looking for a job is itself a job." Marianne Moore, who runs job clubs at the Windsor Unemployed Help Centre, tells people to make sure their resumés have lots of white space. "You need your [grade] twelve" for just about every job in town, she tells them. Even as hopes are ratcheted down in the age of falling expectations, people face the paradox of being subjected to what might be called a "qualification-escalation ratchet" as employers offering even the most mundane jobs are able to demand rising levels of qualification.

Moore is a large women in her early thirties who devotes a lot of attention to fashion—stylish hair, intricate makeup, long fingernails with painted designs. She's both earnest yet street-smart, one of Ed Broadbent's "ordinary Canadians." She worked her way up from being front-desk receptionist at the Help Centre.

When she started she had no training as a group leader, and the first few weeks were "traumatic." Like many other workers, she played it by ear, picking up insights and techniques as she eased her way into the job. She learned a lot from the people in the classes. By now she has a knack for the work and knows the shifting, sliding world of work in Windsor. Like many of the people in her classes, she's struggling to upgrade herself. She already has a certificate for social welfare courses she took at St. Clair College and is heading to the University of Windsor to pick up sociology credits, though she doesn't seem quite sure what good it will do. "It's just gonna be a piece of paper," she says. "It's just life experience that matters."

Is formal training or learning on the job more important? Moore is ambivalent about the training syndrome and the increasing cre-dentialism that accompanies it. "They're both important. But some-how having the life experience and saying 'I can do that' rather than saying 'I've studied that' seems crucial. Unfortunately a lot of employers want that paper. . . . And besides people have to do *some-thing*. You may as well upgrade. Why not?"

Marianne Moore's job clubs attract what she calls "the United Nations—it's unreal." She has a former Iraqi construction contrac-tor, who recently arrived from Montreal with his wife and five kids; a single mother who has not worked since 1987 and recently returned from a marriage in Australia with the hopes of getting work in the health-care field; a Hungarian immigrant with retail experience; a Czech woman, who sits next to the Hungarian; a factory worker from Windsor who just got his Grade 12 upgrade.

Then there is the 7 Eleven clerk who was fired because the boss accused him of pilfering. He's angry at the world, and today his anger is focused on TV comedienne Joan Rivers and her ilk. "When you don't work you wind up watching these shows that are going nowhere," he says. Moore tries to get him to refocus on what they are supposed to be doing, a brainstorming session about networking as a means of getting job leads. "That Donahue is . . . he's another bone-head!"

They sit around in a hollow square formed by institutional tables, their first names displayed on cards. It is a bit like a seminar that a management consultant might put on for a group of business peo-ple, except that in place of a carpeted, catered hotel meeting room the venue is one of the Centre's austere windowless rooms. The only concession to decor is a lonely Lawren Harris poster, part of a series sponsored by the Canadian Labour Congress on art reflecting work-ing-class life.

Vincent is a fifty-two-year-old immigrant from Barbados with years of experience doing repairs and estimating in body shops. The chemicals in the body-shop paint and the dust from the work finally got to him, and his doctor told him to quit while he still could. He wants to get on at a car-rental agency doing something behind the counter. He feels he will have to build on his Grade 10 if he wants to even think about getting something halfway clean for decent pay, because he has absorbed one of the key falling-expectations messages of the Job Club: "Everything is computerized now. The days are over when you could make $20 an hour by walking into a factory without a formal education."

The Job Club is like a self-actualization session or a meeting of Alcoholics Anonymous. Sessions are as much motivational as instructional, as if the human potential movement were in charge of the training agenda. There is the usual go-round in which people introduce themselves and describe their situations and feelings about being unemployed.

"You're good people," Moore says with a big smile, obviously trying to use enthusiasm as a confidence-builder. "Unemployment is just unfortunate this time. You've gotta work on yourself and get out there. Congratulations for coming in. It's Day Three. It would have been really easy to stay home and watch the soaps today, wouldn't it?"

Despite her approach, Moore senses an increased feeling of helplessness among the people who have come to the Help Centre in the past few years. People just don't know what to do, where to go, she says. "They often ask me, 'Am I doing something wrong? I'm just not *getting* anywhere.'"

But the Job Club is not all stroking and self-esteem. Moore also cracks the whip. "There's nobody gonna get you a job," she scolds. "You have to get yourself a job. Your skills have to be very flexible."

Saying this, Moore flourishes a copy of the Windsor Development Commission's manufacturers' directory, the one with the cover decorated with Champion spark plugs, Wyeth birth control pills, and several other products no longer manufactured in the city. "This will be your bible," she tells the class. The Job Club participants learn about the "hidden job market," where 80 per cent of jobs are never advertised. They hear about the importance of networking. A good way to develop networks is to get involved in volunteer work: this will help you build a contact list and get experience as well as boost self-esteem, they hear.

The Windsor Help Centre shares this whole approach with the CEIC, one of its principal sources of funding. The government feels, probably correctly, that a centre staffed on the whole by working-class people and backed by the institutional weight of organized labour (no small consideration in a union town like Windsor) is more likely to succeed in attracting the unemployed to federally sanctioned courses than its own Canada Employment Centre, a colder and more intimidating place. But a different venue does not mean any fundamental shift in the ideological messages being retailed to workers, and these are laid out in *How to Find a Job*, a skilfully designed book with a handshake logo published by the CEIC in the depths of the depression in 1991.

A testimonial to personal initiative, the hundred-page book allows that "it's not easy" to find a job these days. It reminds the jobless that there are "a lot of people like you." You learn that your personal resumé is "your advertisement for you"—it might open the door to an interview, the chance "to sell yourself." The upbeat tone is leavened by stern reminders. "IS YOUR ATTITUDE HOLDING YOU BACK?" Be confident, assertive, and know yourself. Upgrade your skills. Get out and flog them: "The job market is like any other market today: the competition is getting tougher." Remember, say the people who are in a position to cut off your benefits, "*looking for a job is the most important job you can ever have.*"[24]

This particular job may be small comfort to the many searchers—particularly people from the working poor—who are becoming aware of the structural transformation of the job market and know they can only aspire to a marginal job. It is surely a reminder that unemployment is, indeed, *personal* trouble. Appeals to self-confidence and self-actualization build a sense of potential workers being driven by deficiency. In his 1976 book *The Diploma Disease*, Ronald Dore describes the creeping credentialism fostered by the expansion of formal education and training: "Learn this or you will not get the chance to be a doctor or a carpenter; nobody will *give* you a living."[25]

According to Dore, learning easily becomes a means to an end (in this case a certification that may lead to a job), a passport to the land of work populated by "deficiency-motivated" people. Dore cites psychologist Abraham Maslow's description of such a person, a man who "cannot be said to be governing himself, or in control of his own fate. He . . . must adapt and adjust by being flexible and responsive and by changing himself to fit the external situation. He is the dependent variable; the environment is the fixed independent variable."[26]

"The onus is on you—you're responsible for yourself in all of this," Marianne Moore tells the unemployed.

———

One government initiative that the Help Centre backed captures the ethos of the age, and it may serve notice to the future. It is the program aimed at turning welfare recipients into entrepreneurs. Launched under the NDP government's jobsOntario fund, the initial Business Ventures information sessions attracted a hundred people interested in getting into business for themselves.

The image of the mom-and-pop operation has taken on the lustre of job creator. In the mid-1980s a consultant and Massachusetts Institute of Technology lecturer, David Birch, issued a controversial report claiming that small businesses spawned eight out of every ten jobs. In fact, small businesses are volatile. They are assembled and disassembled like knock-down furniture. They hire early in an economic recovery; their activity contrasts to the tendency of major corporations, which lay off late.

"In the service businesses generally," says Mitchell Fromstein, whose firm tracks hiring trends, "there has been a very definite trend that has favoured the large companies."[27] Fromstein ought to know; he is CEO of Manpower Inc, the largest temporary-help agency in the United States.

To run her Business Ventures, Help Centre boss Pam Pons hired Jake Oliver, the former news director of a local radio station. Oliver, who describes himself as "very, very competitive," has watched

Windsor change in the past few years, its "nest-like quality" evaporating. For him the key ingredient in successful microcapitalism is "desire." So Business Ventures is oriented not so much towards providing the poor with capital as towards "building self-esteem, helping them believe that they *can*."

The program culls two dozen likely business people from the hundred who express interest in self-employment and then puts them through four months of sessions on self-assessment, bookkeeping, "time management," communications, stress management, marketing, and financial planning. Each candidate then presents a business plan to a committee chaired by the head of the local labour council and including a lawyer, an accountant, and a financial planner. The committee decides which projects will be funded.

Oliver downplays the importance of access to capital. Risk-taking and the creative juices of the individual entrepreneur are extolled by everyone from mom-and-pop businesses to the chartered banks that dominate the Canadian economy. The banks would surely jump at the chance to put their money where their collective mouths are. Or does the greed that overcame their vaunted probity in the cases of the casino capitalism of the 1980s and especially the Reichmann debacle (generating employment for lawyers and investment analysts forcing the banks to make room on their balance sheets for massive loan-loss reserves) not extend to the capital needs of needy citizens who wish to become capitalists?

Jake Oliver thinks the banks are "overly reluctant" to get involved in his kind of projects, though he adds optimistically that several have cut their service charges to new business. "It's the entrepreneurs who really lead us out of the recessions. Governments realize that. That's who led us out of the last recession, and that's who's going to lead us out of this one. That's why the government has put money into this type of thing." By 1994 the Help Centre's Business Ventures program had helped to launch, among other businesses, Essex County's first cat boutique.

If there is one person at the Help Centre with a finger on the pulse of local employment trends (as well as government policies), it is Executive Director Pam Pons. A vibrant woman in her thirties who drives a red Corvette, she laughs a lot and exudes the civic pride typical of many natives of a town long tarred with a grimy, lunch-bucket image. She is convinced that Windsor is free from the stubborn pockets of hard-core poverty that exist in many other Canadian cities, attributing the egalitarianism to the strong union movement that organized what was a strong manufacturing sector as well as a very high level of per capita contributions to the United Way.

Pons's mother was a "homemaker," her father a factory worker. When she graduated from high school there was still something to the old expression that you could either go to the University of Windsor or the University of Chrysler. As a woman, Pons chose the former option. She began working at the Centre the year after it started in 1977. Back then it was called, perhaps wistfully, the Windsor Full Employment Centre, with its roots in the trade union movement that launched it.

Competition for admission to the University of Chrysler now rivals the struggle to get into Harvard, and Pons is one of the many people convinced that training is the key to the future. She and her staff organize bowlathons, golf tournaments, raffles, and bingos to supplement state funding, and they also get money from the United Way. But the Help Centre's principal support comes from government: a purchase-of-service agreement with the city welfare office, funding through Ontario's labour ministry, and the renamed "Education and Training" ministry, and jobsOntario money such as the entrepreneurship program. "We have a division of closures and downsizing," she says, pronouncing it *closuresanddownsizing* because the inseparable twosome have become so commonplace. "We operate a support group for unemployed workers called HUG— Handling Unemployed Groups—and we do career transitional counselling."

Asked about the Centre's total budget, Pons whips out her calculator and—without reference to documentation—punches a series of figures into it. They add up to $1.786 million. The board members "never envisioned" that the office would grow to its current size, she says. "People should get their heads out of the sixties. We have to redefine what full employment is and unfortunately come up with an acceptable level of unemployment, because there's no such thing as zero unemployment. If you're talking about that, well, it's a pretty powerful drug you're taking."

The Help Centre puts as much energy into coping skills like "anger resolution" and life skills such as basic literacy and helping people prepare themselves for job interviews as it does into specific "hard" skills like those offered in computer classes. Pons has nothing but scorn for the private sector's record in worker training. Like many labour people, she supports the idea of levying a training tax on business, a training levy that would be refunded to firms that actually did run training programs. For Pons, training and coping and a new kind of economy are crucial. Windsor is not going to miraculously get a new megaplant that will put all of the workers back to work, "So we clearly have to diversify." She fully supports the small business program, one of the latest policy fads. Being your own boss gives you more control over your life, she says, and this is a welcome change from the threat of plant closures that worries the entire community.

The idea of people gaining more control over the institutions that affect their lives is as old as the notion of democracy itself, but putting the unemployed to work as small business operators simply means turning private troubles into private enterprise. In a world in which most people have little control over their working lives it may be an attractive idea, but in an economy dominated by a few muscular multinationals, it is simply not a realistic alternative.

It is a comment on the times that a labour-oriented organization like Windsor's Unemployed Help Centre has become consumed with shoring up the band-aid programs of a rotting welfare state. As joblessness grew, the Centre diversified, offering a complete menu of programs to help workers adjust to the new realities of the labour market. This is also, perhaps, a measure of the defensiveness of working-class organizations in this new era. Trade unions have consistently addressed the issue of *unemployment* with arguments about the need for industrial strategy, the need to control both the global strength of capital and the ability of business to deploy job-killing technologies. They have been less able to address the *unemployed* themselves—a group harder to organize even than the new legions of part-timers working at the small workplaces of the service sector.

The contradictions within organized labour are apparent. For instance, in the arena of training and what the training trade calls "adjustment" we have on the one hand Windsor's Help Centre, with its potpourri of programs, many of them focusing on the private troubles of workers. On the other hand we have a similar group in Toronto, the Metro Labour Education Centre, which criticizes the approach that emphasizes "job search techniques, resumé writing, sessions in personal hygiene, correct deportment for job interviews, how to find that hidden job market." According to the Metro Labour Education Centre, "These are the programs favoured by the private sector management consultants typically trotted in to provide adjustment services." In this form unemployed workers get a relentless message: "Here's what you need to sell yourself, to get out there and beat the competition."[28]

This is in part what the training gospel is all about: more private troubles. There is, of course, nothing wrong with adult education, with learning new skills to do more interesting work and participate more fully in public life—or just learning for its own sake. Scandinavian culture, for instance, draws on a rich tradition of adult education often unrelated to particular tasks, and strains of this approach can be found in the history of the Canadian workers' movement.

Mechanics' Institutes, though often sponsored by local notables, were well used by workers who attended lectures and used the reading rooms and libraries at a time when there was little or no public provision of books. In the 1870s craft workers agitating for shorter hours met in the Institute in Hamilton. By the turn of the century Frontier College began providing literacy courses to the men working in mines and forests who produced the wealth that spawned a nation. The Workers' Education Association also provided opportunities to those short-changed by the conventional system. Later, during the Great Depression, the Catholic-led Antigonish Movement used adult "study clubs" to promote community organization and co-operative development in Nova Scotia, where the first credit union in English Canada provided an alternative to outside banks.

More recently, Windsor has seen a resurgence in adult education based in the working class, with the success of the Canadian Auto Workers union in negotiating provisions for paid educational leave (PEL), which allow it to run union-controlled classes for its members. The CAW rebuilt its old property on the Lake Huron shore, transforming a rustic summer camp into a finely appointed Family Education Centre, with training sessions in everything from safety and health to working-class history. Over four thousand people have taken the courses since they began in 1978. The PEL program delivers a blunt, dump-the-bosses-off-your-back message aimed at building "a cadre of activists" with the knowledge and self-confidence needed for effective trade unionism.[29]

Margaret Peever learned about union activism and the personal counselling in CAW leadership training courses she took when she was still working at Windsor Bumper. With her job now long gone, she says that her union training was about as useful as the courses offered by the government. It is not that she has lost sympathy for the union—far from it. Her personal troubles are reflected more broadly. "They are having to struggle, too," she says about the union. "They've lost a lot of really supportive union people who used to give their time. We never got paid for it. We used to volunteer our time and work really hard to educate the public. When you're out of the union and away from it, they don't draw on you the same as they did."

The CAW classes did have some discussion of what happened to members when plants closed. "But that was the end of it," she says, more in sorrow than in anger. "That question was raised in 1987. And look at me. I've got time on my hands because I'm between courses, and there are things I could be doing. They could draw on me, and I wouldn't be looking for money. But they don't do that. They spent big money training me. For what? To sit at home with all this knowledge—talking to my cats?"

She gives out a loud, rich laugh. Even the most strident preachers of the training gospel could not fault her for failing to adhere to it. Peever began to put in a new batch of applications for factory work, including one for a job on the new third shift the CAW finally negotiated at the Chrysler minivan plant. But she didn't get into the University of Chrysler. "I never even got a call."

Margaret Peever's story reflects crucial challenges facing Canadian labour. How can unions maintain a sense of connectedness to people who have been filtered out of the organized, good jobs sector? And, possibly more important, how can they develop that sense of connectedness among the children of those people? As young people struggle for work—any work, from pizza delivery to self-employed computer trouble-shooting—they become harder to organize. There is every possibility that they will not easily identify with the mainstream of organized workers. Who will express the needs of this new breed of workers? Or will their needs simply take the form of private troubles—hidden away far from public view, out of the realm of public debate?

7

COLLARS OF MANY COLOURS IN THE LIMESTONE CITY

As we have done with natural resources, we exploit our human resources on a narrow, short-sighted basis.
– Special Senate Committee, *Poverty in Canada*, 1971

I F WINDSOR has its annual gala cross-border Freedom Festival extravaganza, each year now, when the lilacs blossom and the university students depart, Kingston plays host to a job fair.

The local Kingston Area Training Advisory Committee has rented the main ballroom at the city's biggest hotel, the Ambassador, just on the outskirts of town. The unemployed from both the city and outlying Frontenac County are invited in to sample coffee and donuts. People wander tentatively around colour-coded displays: blue for social service agencies, red or yellow for public educational institutions, green for employment services outlets, orange for private trainers.

One booth colour-coded orange has a huge poster of a tattooed Axel Rose, lead singer of Guns 'n Roses, his image captured in full voice, his mouth roaring into the mike. The caption reads: "What I tell any kid in school is 'Take business classes, whatever else you do.'" The master of ceremonies at this Stepping Stones to Employment conference is an extroverted social worker who accompanies himself on blues harmonica:

"I don't mind learning everything you've got to teach
I'll keeping peeling off the fuzz 'til I find the peach
Job, job, job
Don't you jive me now
Job job job
Gotta try somehow
Job, job, job,
Broken record on the big machine . . ."

"This is an employment fair," the social worker shouts. Putting aside his harmonica, he explains to a curious crowd that the employment game is all about playing the "angles." It's the job market as billiard contest. "What we're trying to tell you is that these are the people in your community who have some connection to the concept of employment."

Officials from the Training Advisory Committee nod in agreement. When the current training push began to build momentum in the late 1970s, such advisory committees emerged in many cities. Although the idea came from above—the government—they were still called Community Industrial Training Committees (CITCs). With employer-sponsored training all the rage at the time, both Ottawa and Queen's Park were making another of their periodic attempts to prod the private sector with enough cash that the companies would begin to actively train their employees. Conceived as a tripartite, corporatist initiative, the CITCs were composed in the main of local business people, chambers of commerce, and educational authorities, with minimal labour participation. According to an internal policy document, the success of employer-sponsored training, or EST, is "entirely dependent on the response of the private sector."[1]

Nonetheless, the CITCs had a life of their own, sustained by state funding and the mounting interest in training as an answer to the vexing questions of joblessness. By the time of the gathering in the Ambassador Hotel ballroom, the NDP was in office in Ontario and the Kingston Area Training Advisory Committee was making its best corporatist efforts to include labour, not to mention disabled people, women, and ethnic minorities. Advisory Committee head Donna Miller, the principal organizer of the job fair and a strong believer in "industrial input" into public education, says it is wrong to call what happened in the early 1990s a "recession." Rather, it is merely "industrial restructuring."

At Stepping Stones to Employment, people shop around at the supermarket of training possibilities. The Kingston General Hospital offers vocational assessment. The Computer Coach hopes to sell courses conducted by "professional but friendly" trainers. Kingston & District Immigrant Services provides help to newcomers. The welfare and UI offices have representatives. So do every continuing education, literacy, and counselling service in town. The NDP government has created something called jobsOntario training (they must have hired an ad agency to think up the name) that matches the poor to "newly created jobs in the private sector that have a training component."

The largest recipient of jobsOntario training money in Frontenac County turned out to be a hot new franchise restaurant chain, as much of a concept as a place to eat. East Side Mario's bills itself as "an Italian American Eatery." If you ask to speak to Mario they look at you like you've lingered too long at the happy hour. By the end of 1993 jobsOntario training had created 190 jobs in Frontenac County: one in four jobs was in the hotel and restaurant business; 17.4 per cent were in retail trade; 12.6 per cent in "other services." Clearly, government training dollars were being funnelled into the low-wage end of the labour market, where the most precarious employment is being generated. Manufacturing industries took up funding for 4.2

per cent of the jobs generated. Health and social services came in at 3.1 per cent. For jobsOntario purposes, a thirty-hour week is defined as a full-time job.[2] At $10 an hour (a high rate for waiters and many other workers in the female-dominated personal-services sector), this translates into an annual wage of $15,600, before taxes.

Brian James Wilson of East Side Mario's Shason Foods Inc. sums it all up: "In these tough economic times, small businesses such as ours have many front-end expenses and start-up expenses and the recapturing of some of our training costs from the jobsOntario program will assist us in our profit-loss situation."[3]

Like the booths at most conventions, the Stepping Stones to Employment get-together offers participants the chance to attend smaller sessions. In the Ambassador's Lisbon Room, a former maintenance mechanic at the Alcan plant is telling seventy people how to develop a positive mental attitude. Herb Waldie, an evangelist for the human potential movement, asks the crowd to call out the characteristics of "a winner." Individuals in the group offer up a variety of responses. "Enthusiasm," someone yells out. Another says, "flexibility!" Once the ball gets rolling we get "risk-taker," "people skills," "team player," and "self-confidence."

"There are two things we need to be successful," Waldie explains. "Attitude and skill." He says all of the characteristics of a winner that came from the group depend on attitude. "At least 80 per cent of our success is attitudinal. . . . I used to be one of those people who would bitch and grump. But we need to become reverse paranoids, to convince ourselves that everything bad that happens to us has a positive outcome."

Waldie has a binary perspective. For him there are obviously two sorts of people in a gambler's society, because his entire approach is framed with references to "winners" and "losers." He tells the attentive group about a recent conversation with a manager of the Goodyear plant in neighbouring Napanee (a plant with no workers, only "team members"—and no union). Apparently the boss of the tire factory told Waldie that when they looked for new employees the search was 99 per cent attitude. Waldie says that in the same way people have control over their health when they eat well and avoid smoking, so too can they also gain control over their own lives.

He then shows a video by his mentor, a successful car dealer. "If it's to be, it's up to me," says Brian Tracy, whose video proceeds to warn the Kingston poor about "the valley of excuses." The tape is more explicitly ideological than Waldie's personal presentation. Tracy attacks government and the "urge" to collectivism and socialism. "You are free to choose," intones the voice from the video monitor. "You determine the quality of your financial life."

Although attitude adjustment is never far from the top of the training agenda, the Kingston Job Fair is not entirely given over to the banalities of sturdy individualism and laissez-faire. Stepping Stones to Employment also offers the chance to sign on for some "hard" skills. A half-dozen private trainers offer courses focused mainly on computers. "*Learn quickly to earn quickly,*" says the Metzler Business School, one of the old-fashioned secretarial schools that for decades have marketed their services to young women. A fixture in Kingston for half a century, Metzler offers basic accounting, business math, and business English as well as the usual computer packages. "In an effort to create a business-like atmosphere, a dress code will be in effect for Business Administration courses," says a Metzler promo piece.

A more sophisticated operation is the Career Development Institutes Ltd., a blossoming national training business spun off the instructional wing of the Control Data Corporation. CDI operates over twenty training outlets from Halifax to Victoria. With its slogan ("The Power to Succeed") and its go-getting staff, CDI represents the vanguard of a new private-sector training industry. CDI founder and president Bruce McKelvey, a former Control Data executive, has sat on government training advisory boards set up by both the Mulroney Conservatives and Bob Rae's Ontario NDP government. He has examined the German vocational training system on behalf of the former Ontario government of David Peterson and is a veteran of boards of trade and university computer advisory committees.

The CDI booth at the Kingston Job Fair is staffed by several well-dressed sales reps who know their business. They hand out copies of *Partners*, the CDI publication. This issue has a picture of McKelvey with Manitoba Minister of Education and Training Rosemary Vodrey "reviewing MTTC's partnership with CDI." (CDI was brought in to run the Manitoba Technical Training Centre in 1983.) A front-page headline reassures us that "There Are Jobs Out There," and the publication explains the importance of flexibility and of being responsive to business needs.[4]

A potential client hesitantly approaches the CDI booth and asks the salesman about the cost of courses. "There are funding sources available thru UI, OSAP [Ontario Student Assistance Plan], or whatever," he quickly replies, "but that's the last thing we look at." CDI's Kingston branch is more highly capitalized than the old-style business-colleges-cum-secretarial schools of the Metzler variety. CDI is heavily computer-oriented. It is poised to take advantage of what it sees as a new business commitment to training, anticipating that companies that contract-out other functions will be doing the same with employee training. The training market also trades on the anxiety many Canadians feel about being left behind as computer applications change how work is organized and carried out.

Then there is the government. Because training is seen as a key solution to unemployment, federal and provincial governments have been pouring money into the area. Ottawa has spent money diverted from the UI fund. A typical CDI ad in the help wanted section of the Kingston *Whig-Standard* offers "Tuition Free Training!!" funded through the Canada Employment Centre and available to UI and welfare recipients as well as the general public. CDI sells the government courses in "network installation and maintenance, advanced software applications, PC maintenance and troubleshooting."[5] The company has grown from having five Career Colleges in 1987 to twenty-five in 1994, by which time it also had five business training centres and a Career College in Beijing.

The CDI's office in Kingston's Peachtree Plaza is run by Rick Lawless, with a staff of five instructor/sales reps who work in a roomy, carpeted space. Students have their own terminals, networked into a central system. Lawless reckons the average student is about thirty years of age. He is an enthusiastic fellow whose favourite word seems to be "exponential," especially as applied to the rate of workplace and social change being wrought by the computer. Lawless agrees that CDI is doing some of the same work that the old-style secretarial schools used to do. But it was well poised to expand when the business sector started "buying computers as fast as Heinz makes pickles"—this at a time when the business colleges were still selling courses that featured IBM Selectrics. The Kingston branch opened in 1993, but earlier on the firm had run computer maintenance courses at the Collins Bay Penitentiary "in partnership with Corrections Canada." (CDI-talk, I find, captures the most up-to-date jargon. Customers become "partners." Students are told of the importance of "accepting change," or "When you have the right skills, the sky is the limit.")

According to Lawless, fully 30 and 40 per cent of his business comes from government seat purchases. The rest is private, like the thirty-three doctors who took a CDI course after the provincial health insurance plan demanded that they submit their billings electronically. The doctors wanted to know enough about the software so that they would not be dependent on the skills of their office workers. "Training," Lawless says, "is big business now." The company is the prototypical flexible operation so beloved of the management theorists and gurus of the 1990s. Lawless says CDI can adapt rapidly to shifts in the market—a market that in this case is often supplied by the state.

"The government might say, 'Oh gee, we've got $10 billion we're gonna spend on training.' At which point they may pick up the phone and ask if we can take a hundred students, and we say 'Yes' and rent some space. The beauty of our head-office support is that I can have a valid program for fifty people up and running in about a week."

Everything seems portable in the information age: jobs, skills, training facilities. *Place* no longer seems to matter very much. Things are intangible. "Knowledge," insists Lawless, "is the product." He explains:

- how employers are looking not for people with narrow skills, but for workers who can run not one but several programs and also know how to take the hardware apart. ("A half-dozen people like that and you don't need twenty-five secretaries.")
- that government simply has to get together with business and the training industry or Canada will be in "the stone age" compared to Singapore, Japan, and Germany.
- how employers see computerization saving them big money. "The more things that machines can be made to do, the fewer people they have to hire," he says. "Anyone who makes the commitment to learn" about the technology "is probably going to be employable."
- that private-sector instruction is more flexible and up to date than public education.

For Lawless and many other pragmatic-sounding optimists, the shift to computers parallels the shift from the age of the horse-drawn cart to the age of the automobile—with the resultant demise of the famous buggy-whip manufacturer. "If you couldn't learn how to be an auto mechanic, maybe you didn't have a job for a while. Henry Ford saved a lot of people's bacon by hiring them to work in his plants. But they had to be retrained. There is going to be a rather painful transition. I'm sorry if you don't want to go back to school or learn how that machine works, but if you want to work you're going to have to."

Such reasoning presupposes a number of conditions. Not the least is that workers had to be retrained for work on the assembly-line. In fact, the line where Henry Ford's robot-like human manikins toiled was designed to remove as much skill as possible from the work. Much of the "retraining" that accompanied the move into the cavernous factories of mass production involved either cajoling or forcing formerly independent craft workers and farmers to submit to the discipline imposed by a new class of industrial engineers. In turn these engineers, in the employ of the newly emerging megacorporations like Ford, contrived systems that forced workers to do particular kinds of work in exactly the same way, over and over again, day in and day out. This is not skills training in any meaningful way; rather, it is a form of cultural training designed to moderate human behaviour and expectations, adapting people to the needs of employers. ("Accepting" and "welcoming" change is a key to success, the CDI says.) The goal of these firms and their industrial designers was to deskill and fragment the work as much as possible. What the new industrial workforce had to "learn" was to accept the rhythms of machines they did not control and accommodate themselves to externally imposed alterations of their working lives.

As Bertrand Russell pointed out in his essay "In Praise of Idleness," there are essentially two kinds of work—moving matter around on the earth's crust; and telling other people to do so.[6] This is changed not one whit by the fact that much of the new world of work involves the moving around of intangibles such as electronically stored information.

A second presupposition is the reasoning that compares the historical transition to a manufacturing age to the emergence of today's information-based service economy. This comparison contains the implicit assumption that the job opportunities in the new economy will in fact emerge. Capitalism is without doubt a dynamic agent of change, but with the decline of employment in manufacturing and the automation of so many services, there is no obvious take-up sector to pick up the employment slack.

A study conducted at Queen's University revealed that the proportion of employees who use computer technologies on the job in Canadian business tripled between 1985 and 1991, rising from 13 to 37 per cent. CDI is apparently onto something in aiming its pitch at training people for the new world of office work. But it would do well to examine *who* is being trained. The Queen's study also showed that the dynamic services, where the future of work is said to rest, are also reproducing the familiar pattern of employers training managerial and professional personnel. Between 1985 and 1991 a dramatic shift occurred in the profile of trainees, with fewer clerical workers learning about the new technologies and more high-level employees getting training. Bosses need bosses. Furthermore, although women make up 43 per cent of the workers using computer-based technologies, they received only 35 per cent of the training.[7]

But what about all the skilled work involved with computers—the world of systems design and software writing? This is a seemingly exotic place with inhabitants who speak mysterious languages. It has jobs, some of them good jobs. In Kingston, Rick Lawless tells of a young man who came into the CDI sales office searching for a thousand-dollar programming/analyst course. The former St. Lawrence College student complained that the three-year computer studies course he had just completed had not offered enough time on the machine. "The rest of it was English and basket-weaving," Lawless says, exhibiting a fair amount of scorn for the public education system and its idea of job relevance. The beleaguered student had apparently wanted to get "current, leading-edge training in UNIX operating systems and COBOL" at St. Lawrence but "got none of that."

According to Lawless, the local community college is using ten-year-old software, but he also says that the government throws money at the colleges with no "quality control." When he visited the place, he says, "The whole operation was run like a Bolivian customs agent. That kind of delivery of a product is going to hurt you eventually."

Indeed, St. Lawrence has been hurt. In 1994 government cutbacks forced it to announce the layoff of 15 per cent of its staff. As usual, it heralded early retirement as a method of cushioning the blow. The college is where working-class Kingston goes to keep up with the life-long learning revolution, and enrolment dropped as a result of the public-sector cuts.[8]

The growing importance of firms like CDI and the waning of

public facilities like Kingston's St. Lawrence College are typical of a national—indeed, continental—trend. In the United States private education companies have expanded dramatically since the 1970s. One outfit, Phillips Colleges Inc., is a chain that operates ninety-one outlets tapping into government training schemes.

In spite of the record of Canadian business in worker training, Ottawa has been steadily redirecting money from public-sector to private-sector training. Clause 13.03 of the Canada-British Columbia Labour Force Development Agreement states, "Canada intends, over time, to gradually decrease [the] level of training given by public authorities . . . and to promote, through increased funding arrangements with private sector organizations, an increased role for the private sector in training."9

This logic promotes a training market in which public and private operators compete. Private operators like CDI may have no wish to provide broad programs like "English and basket-weaving," preferring the more lucrative—and narrower—chunks of learning demanded by employers. Once Canadian training is privatized, there is every possibility that U.S. training chains like Phillips Colleges will be able to move in and take over. Because of the provisions of the North American Free Trade Agreement, such outfits will be immune from any Canadian protectionism of the domestic "market."10

Kingston's Employment Services Branch is housed in a 150-year-old limestone building just down the shore from the old Kingston Pen. In the waiting room of the city job office there is a corkboard with the notice, "To Our Clients: Please Read," and a handful of clippings pinned to it. One piece quotes a management consultant: "People are commodities now. When you become useless, you're out the door."

Meghan Charters, an Employment Branch caseworker, probably knows as much if not more about the local labour market than anyone in Kingston. She has nine years of experience trying to help welfare recipients find work and cope with their situation, and her "Life

Skills Coach Certificate" allows her to instruct other group facilitators on how to conduct sessions in anger management. The people she sees "are fighting profound discouragement." Many have already been through the training cycle several times. Others have been through somewhere between two and seven "career changes" in five years.

In the end, Charters says, it all comes back to training and increasing worker qualifications. Given the mobility of capital, its lack of any roots or loyalty to location, people will have to chase jobs, keep upgrading, "be mobile." But she displays little of the upbeat individualism so evident at the Stepping Stones to Employment job fair. Even though she subscribes to the prevailing orthodoxies about the need for competitiveness, R&D, and the future being tied to information-based industries, she also shares with the working-class people she advises an overriding concern for the demand side of the labour market equation.

In Kingston, with its overwhelming dependence on public and parapublic employment, the demand for labour has slackened as deficit-obsessed governments slash away at their costs. *Growth sectors?* A training industry is growing up everywhere, "feeding off where our economy is shifting."

In the lobby of the Employment Branch Services there is a report from the Federal Business Development Bank, and in large letters on the cover page one of Meghan Charters's colleagues has scrawled "Success Stories." The report features thumbnail profiles of "outstanding young entrepreneurs." One of these entrepreneurs set up a chain of funeral homes. Another became a supplier of specialized software to video stores. One established a service that records your marriage vows on videotape for immediate playback on a giant screen during the reception. Still another operates a new firm that bids for contracts on the privatized management of provincial parks.

Where can people look for work in Kingston? They can forget about a solid job with the city's biggest employer, the military base, Charter says. "I tell my clients, 'Don't look for permanent work there. Forget it. It's all on contract.' The base is a big employer but not a stable employer." At the start of 1994 Corrections Canada told Charters that they were hiring in Kingston. "There's definitely some growth there," she says.

Indeed, the "corrections industry" is one of Kingston's biggest employers, and it is a growth industry that is gentle in its touch on the physical environment. "Corrections is a clean industry," the regional head of personnel for the jails says. "No big smokestacks."

As defined by Daniel Bell, the kind of society represented in Kingston (particularly by its leading planners and the ideologues of the day) is quintessentially postindustrial.[11] Bell wrote his influential book about postindustrialism in 1973. It was heavy on the importance of innovation. At the same time a different take on those developments also appeared, from a thinker who did not distinguish the modern from the postmodern. Ivan Illich wrote: "Periodic innovations in goods or tools foster the belief that anything new will be proven better. This belief has become an integral part of the modern world view. It is forgotten that whenever a society lives by this delusion, each marketed unit generates more wants than it satisfies. . . . New models constantly renovate poverty." The conclusion, according to Illich: "The underprivileged grow in number, while the already privileged grow in affluence."[12]

This trend is readily apparent in Kingston, where, for instance, a giant chain store with a huge ugly sign screeching out "Blockbuster Video" threatens to drive the smaller locally owned outlets into the ground. When Bell and Illich wrote about our fast-paced era, no one had even heard of video stores. One of the town's most successful restaurants and a weather vane of downtown Kingston's gentrification is the limestone-walled Chez Piggy. It is housed in a building that used to be a stable and attracts an upscale clientele based in the main on the Queen's professorial-scientific crowd as well as the other professionals who can afford its prices. In the summer its handsome courtyard patio is a magnet for hordes of tourists. Some have heard that the place is co-owned by one of Kingston's better-known citizens, Zal Yanovsky, former television producer and before that guitar player for The Lovin' Spoonful. For all that, Yanovsky is alarmed at the big tourism push, describing it as "very dangerous" because it overlooks year-round citizens and makes large concessions to whimsical outsiders who may or may not come. But Yanovsky's localist sensibility is rare among businesspeople here, particularly the tourism-obsessed downtown merchant class.

He sees a town divided, as it has always been, but maybe more so. Yanovsky makes a point similar to Ivan Illich's, in slightly different terms. He feels there is not much of a local middle class and that instead there are, on the one hand, what he calls "upper middle class" people making $35 an hour and, on the other, people concentrated in a receding, welfare-based north end. "That whole other section of Kingston is a poor white-trash ghetto," he says. "They're fucked. Third and fourth generation. Queen's does little for them. The gentrification—the last thing you want in your beautiful gentrified downtown core is a bunch of yokels. They live further and further away. There's very little exchange—economically, intellectually, philosophically—between those poles."

Queen's University is one of the institutions that has done much to shape Kingston. An eminent Queen's economist and former principal, John Deutsch, the first head of the Economic Council of Canada, wrote—before the gentrification and before Chez Piggy—about how rooted Queen's and Kingston are in each other. There was, he said, an intimately symbiotic relationship between town and gown. "Even in these times of social terminology and economic sophistication," he concluded, "the sharing and mutual interests are past measurement and analysis."[13]

The relative importance of an elite university like Queen's becomes more acute with the decline of manufacturing industries. Universities are crucial to the postindustrial worldview. Ever since Daniel Bell helped to label this new era, a thousand futurists and pundits have gone on to produce an endless stream of wide-eyed predictions about the knowledge or information-based economy, in which knowledge factories like universities occupy key positions.

Equally important are efforts at research and development (R&D) that result in new products, a process usually described as "technol-

ogy transfer." This can take place either in the university labs that
are under increasing pressure from budget-conscious administrators
to commercialize their research, or in corporate R&D centres. The
local Economic Development Commission is so convinced that the
future of this postindustrial "21st-century community" lies in
"breakthroughs in synthetic and advanced materials" that virtually
its entire pitch for investment is based on R&D: traditional working-
class brawn is replaced by "mainframe muscle" in an "executive
environment."

The language used is almost a caricature of itself: "collars of all
colours . . . leading edge . . . old stones turned to new purpose . . .
harnessing technology . . . tomorrow's products on today's drawing
boards . . . biomaterials to replace nature's originals."[14] Such talk, in
Kingston at least, is more than the usual hype of a town anxious to
attract new investment. The place is said to be one of the best places
in the country to live. It really is pretty, clean, and well located. Its
scale is human. Its amenities may indeed attract highly skilled
researchers.

In 1993 the local Alcan lab won the National Quality Institute's
"Award for Business Excellence" in invention, having produced a
strong, light-weight aluminium foam whose fire-resistant qualities
promised wide construction and automotive applications: just the
sort of breakthrough that planners relish. "You could call it the Stan-
ley Cup for the research and development set," gushed *The Whig-
Standard* under a front-page colour picture of the foam's inventors,
Iljoon Jin, Douglas Kenny, and Harry Sang. The award "points
towards Alcan's ability to create new technology that can be spun
into new sales and new jobs."[15]

Within a few short weeks of the award's announcement, the bub-
ble of civic optimism collapsed with the dull pop of corporate reality.
Alcan had an announcement of its own. The corporation was cutting staff at its Kingston R&D lab, eliminating forty-four jobs.
Within a year twenty-one more positions would disappear, for a total
staff reduction of 37 per cent. Apparently there is no automatic connection between successful innovation and "new jobs."

The notion of postindustrialism hinges on the type of society that we are moving away from; it is inadequate for showing where we are headed. And what is happening in Kingston and Windsor and so many other communities is that collars of all colours really are being traded around. In the shift from a goods to a service-based economy, new varieties of class configurations are being *forged*, to use a metaphor from the smokestack era.[16]

If the future depends on figuring out how to manipulate genes to produce bovine growth boosters aimed at developing the latest generation of supercows and adding new wrinkles to basic fruits and vegetables, then a development such as the new Biosciences Centre at Queen's is significant. (But bio-tech is not a big job creator. According to Richard Barnet, the largest bio-tech firm in the United States employs only 2,639 people.)[17] So, too, is the shift in the direction of a university that, although overwhelmingly dependent on public funds, has been the first in Canada to "privatize" one of its programs, quadrupling the fee it charges for its MBA program to over $20,000. But it is unlikely that most Canadians will find jobs as geneticists or top business administrators. Far more likely, if present trends continue, is a career in one place where the demand shows no sign of letting up: the social casualty ward.

———

Daphne Klassen-Hayes hopes to find a security job in the social casualty ward. Correctional services is one of the pure growth industries in postindustrial North America. One in five Correctional Officers is female.

A single parent, Klassen-Hayes grew up in Kingston. Her parents were professors at Queen's. After graduating from high school she married and moved to British Columbia, returning with her two children in 1984 after her marriage broke up. She immediately enroled in a beautician course at St. Lawrence College, working as a part-time salon assistant and doing cuts and perms at home. Then she decided there could be a better future in health care than in hair, so she took out a student loan and enroled in the two-year Registered Nursing Assistant course at St. Lawrence College, where she specialized in burns and plastic surgery.

She soon realized the difficulty of reconciling the demands of a regular hospital nursing career with those of single motherhood. Lacking the seniority for a day shift, she was forced to take various non-hospital jobs, working for a private-sector home-care operator and, later, running the office for a group of family practice physicians. The pay in those jobs was about half of what she would have received as a nurse in a public hospital, but at least she could schedule her own hours. During the four years she spent as a nurse Klassen-Hayes took courses in palliative care and oncology and began to train as a volunteer in crisis intervention, helping the police do follow-up counselling with the victims of major accidents.

"Everybody needs to keep upgrading," she says, seated in the dining room of her suburban home not far from the Collins Bay pen. She tells her story matter of factly, with an articulate self-confidence. She harbours no resentment about the treadmill she's been walking; nor does she boast about her energetic approach. "Wherever I end up I'll have to get more specialization. There's so much pressure you have to have something that makes you stand out. Plus there are all those Queen's students with all those letters beside their names."

Klassen-Hayes needed a better-paying job, but with continuing health-care cuts she decided that there were "no jobs" in nursing. "The health care system in Kingston is in trouble, at least for nurses." She wanted to stay in Kingston, regarding the option of moving to Toronto with the same horror she reserved for the American option that many of her classmates were considering. U.S. headhunters were scouring the town for nursing talent. In 1992 the hairdresser and nurse decided to return to St. Lawrence College for the third time. She realized that if there was a shrinking future in caring for people, with a growing sense of fear of crime there might at least be a place in guarding them.

Kingston has been a prison town since the 1830s, and St. Lawrence offers a two-year course that trains people to deal with convicted criminals. Given high unemployment, prison officials can now demand steep qualifications for jobs that were at one time scorned as being among the most distasteful in society. Daphne Klassen-Hayes enroled in the Correctional Worker Program.

In the old days the career path to employment at Kingston's prisons was well worn. It proceeded either through families, with several generations of a single family commonly working as guards, or along that other institutional road, on which the familiar world of uniforms, chains of command, and authoritarianism of life on the town's military base merged almost seamlessly with that of the prison. Literacy was not as important as being hard-nosed. Those were the days when jails were not yet correctional facilities, guards were not yet correctional officers, and prisoners not yet offenders.

Although prisons are like death in that we tend to avoid acknowledging them or even speaking much of them, things have changed, at least in the types of people the government hires to work in them. There is much talk of "psycho-social" skills. Corrections Canada, the body that runs Kingston's jails, has a corporate Mission Statement: ". . . human relationships are the cornerstone of our endeavour." According to Bob Fisher, the head of career management for the federal prison system in Ontario, this emphasis on social skills means that the "screws" of old have now become "professionals." It is a buyer's market, he says. The applicants who get in must have "either or both, and usually both" postsecondary education related to corrections—university social sciences or community college law and security training—or "experience in dealing with people in crisis."

The point is not that Kingston's prisons have become counselling centres in which concerned staff spend their time nurturing human relations, coaxing convicts back into society. They have not. The jails, from the cavernous old Kingston Pen in the historic prison town of Portsmouth Village to the Prison for Women up the street and Millhaven and Joyceville just outside town, are by all accounts nightmarish places to inhabit and not much better to work in. Suicide, drugs, violence, and murder are commonplace. One can be slashed for glancing into the wrong cell, maimed for muttering the forbidden epithet "goof." Prison workers have high levels of stress, with alcoholism and mental breakdown not uncommon. Canada's billion-dollar-a-year penitentiary system is becoming seriously over-crowded. For many years the number of inmates rose at a rate of 2 to 3 per cent annually, but in 1993-94 the prison population jumped by 8.5 per cent over the previous year. Suicides rose from a count of eleven to twenty-four. While working at Kingston's Bath Institution, psychologist Lois Rosine found that 17 per cent of correctional work-ers experienced effects severe enough to be diagnosed as suffering from Post-traumatic Stress Disorder, "significantly higher than the 1% level found in the general population."[18]

Given this, it is striking that so many people are lining up for the training that might give them a chance at a job in these places. In 1992-93 the Corrections Canada personnel office in Kingston received three thousand applications for what turned out to be about one hundred job openings.[19] Of the total Kingston-area staff of 2,624, 42.6 per cent are guards, or correctional officers. Beginners get $27,478 a year. The top rate for an experienced officer is $38,743.

When Daphne Klassen-Hayes was admitted to the Correctional Worker program at St. Lawrence College, there were six hundred ap-plications for the forty-seven seats in the course, a demand that prompted the college to increase places the following year. Stuart Payne, one of the course instructors, attributes at least part of that in-terest to the new attractiveness of job security and the benefits pack-age of a union job, particularly during a time of high unemployment. Canada spends $1.9 billion annually on adult jails, and prisons are ex-panding at a time when virtually all other government operations are under the knife.[20] All the let's-get-tough-on-crime talk is a boon to the jail business. "It's not like a motel that shuts its doors when it's full," says Payne, a fourteen-year veteran of guard instruction.

The director of the Correctional Worker program is Megan Way-Nicholson, a long-time "case management worker" (formerly known as parole officer) who quit her previous job when her caseload began to approach nearly two hundred files. She says that part of her cur-rent job is to teach her students coping mechanisms, how to deal with stress and the potential for on-the-job burnout that is becoming ever more acute with overcrowded jails and rising caseloads for probation officers. "You have to cope with being spat on, because these people don't want to see you and may well hate you," Way-Nicholson says. Of all the teachers in Kingston, she is one of the very few who screams obscene insults into the faces of her students as an integral part of the curriculum. "It's hard to be both helper and keeper. One day a person will want you to be a helper, and the next day the same person will hate you for being a keeper."

One of the reasons that Daphne Klassen-Hayes made the switch from nursing to corrections was so she could still "work with people." She had been under the impression that this was part of nursing, but the funding squeeze that resulted in the decline of nursing jobs also meant more pressure on those who are working, something reminiscent of the speed-up process familiar to assembly-line workers. Hospital nursing work is increasingly mediated by computerized control systems. Klassen-Hayes feels that nurses do not have enough time to work with people. "There's too much work to do. You don't get a chance to use your human skills."[21]

With her existing training and volunteer enthusiasm, she seems well positioned to find a job somewhere in the social casualty ward. She was among the 4 per cent of applicants interviewed for a job guarding prisoners at a provincial Detention Centre in neighbouring Napanee. She hopes her ability to stay calm and the crisis management skills she learned in nursing will enable her to survive the hard cases. She also knows that to get this kind of a job she needs to do more than just excel academically and get "that piece of paper." She needs a strong resumé with a lot of volunteer work, something to give her an edge. "There's so much pressure you have to have something that makes you stand out." She sees a growing demand for people with crisis intervention skills—even in the local school board, which is hiring correctional workers to counsel young people who do not or cannot fit into the education system.

Another student, Jim Goodwin, whose father was a truck driver for over thirty years, has been working since he was fifteen. He supported himself, his wife, and four children with government loans when he went through the correctional course at St. Lawrence. Those debts are still outstanding, and he is not optimistic about paying them off quickly. He works twelve-hour days driving cab. The rumour he has heard is that he is only one of 1,050 people whose names are on the waiting list for jobs in the Ontario region of Corrections Canada. Even if he gets the call, Goodwin figures he is still two tests and two interviews away from the in-house training course that will qualify him for work at one of the Kingston prisons. Goodwin knows families who have worked in the prisons for three generations, but things have changed since the days when you could get into the system if your father was a guard.

"At one time prison guards were looked on as scum," he says. "The bottom of the list. Now it's one of the most sought-after jobs. A full benefit package, the dental plan. It's a big thing when you have four kids."

Unlike Daphne Klassen-Hayes, Goodwin fits the stereotype of the prison guard. He is a burly man who seems game, seems ready to mix it up with the tough cons in the joint. Like his father, he has a licence to drive the big rigs. He has also worked out west laying interlock brick in Fort McMurray after a stint as an oil-rig roughneck. Returning to Kingston, he got a job on the line at UTDC, a company that made mass transit vehicles. When UTDC shut down, friends convinced him to apply for the corrections course even though he lacked the formal Grade 12 qualification. Goodwin got in as a mature student and graduated two years later, even though he found the criminology, law, and psychology courses tough going.

Goodwin works as a volunteer firefighter. He figures this will look good on his resumé. He hopes his solid work record and formal qualifications will help him get a prison job. He's even ready to put in an application to another region of the corrections department, although his wife, a Kingston native, is reluctant to move away from home, friends, and family. In the meantime Goodwin himself is reluctant to take up long-haul trucking, with its 2 a.m. to 8 p.m. shifts. Besides, the free-trade shakedown has resulted in the closure of four Kingston-area trucking terminals in the past five years as companies centralize operations to cut costs.

"There's no industry any more. Kingston has nothing to look forward to," reckons Goodwin, recalling the twenty-five hundred people who lined up in 1989 for four hundred jobs when Goodyear announced its new tire plant in Napanee.[22]* He knows that the information-era service jobs have nothing to offer him. "There were 350 of us at UTDC and we all went to jobs we never wanted. Taxi drivers, couriers. But there's no money in it." Though he knows prison work is a high-stress job, full of dangers, Goodwin is still keen to get his chance. "It's not like one of these jobs that might last four or five years," he says. "Once you're in, you're in for life."

Technology, it is said, is responsible for much of the displacement of people from jobs. Unions set up "tech-change" committees to deal with the dizzying array of innovation wrought by the microprocessor. The attraction for managers is obvious, as old as the factory system itself. Machines are cheaper than people. They work tirelessly. They want neither overtime pay nor time off. Unlike humans, their tiniest movements can be easily controlled, in theory at least.

Take the case of Garrie Manser, the Kingston spray painter who lost his job when K-D Manufacturing shut down and moved across the river to upstate New York. Had their company not been undercapitalized, Manser's bosses could have saved money in another way, by investing in robotic paint sprayers like the ones that are displacing workers in car plants. When robots do the spraying they use up to 30 per cent less paint than even the most experienced painters like Manser. The reason? They are far more controllable.[23]

Technology is not just about getting things done faster and cheaper. It is also intimately bound up with controlling *how* things

---

* Later, after the plant's opening, Goodyear made it known that it was looking for 130 more workers and got 2,700 applications.

are done. Experimental physicist Ursula Franklin calls this "design for compliance." Her career interest in the properties of metals and alloys led her to an analysis of how the Chinese had developed techniques for casting bronze over two thousand years ago. They used an intricate division of labour that produced the exquisite ritual cauldrons of the Shang Dynasty. Because this work involved an elaborate system of command and control, Franklin calls it "prescriptive technology," which she juxtaposes to "holistic technology." Prescriptive technology was (and is) the preserve of the organizer, boss, or manager. Holistic technology remains under the control of the artisan or individual worker.[24]

For Lewis Mumford the first "megamachine" was not physical at all, but bureaucratic. The pharaohs of ancient Egypt were able to organize the construction of their pyramids because they used a highly centralized machine composed of the labour of a hundred thousand people who were interchangeable, specialized, and controlled from on high. Only a tiny elite had a say in the design of what became one of the world's greatest tourist attractions. But, Mumford wrote, "The workers who carried out the design also had minds of a new order: trained in obedience to the letter, limited in response to the word of command descending from the king through a bureaucratic hierarchy, forfeiting during the period of service any trace of autonomy or initiative." These ancient workers "would have felt at home today on the assembly line."[25]

French psychologist and philosopher Michel Foucault brought this analysis closer to an enlightened modern world when he examined how French schools, military institutions, and prisons of the seventeenth and eighteenth centuries evolved detailed structures, complete with complex hierarchies, stern drills, and detailed record-keeping. Discipline would mould "docile bodies." Foucault showed how regular exercise, training, and work were a recipe for domination. "Historians of ideas usually attribute the dream of a perfect society to the philosophers and jurists of the eighteenth century; but there was also a military dream of society; its fundamental reference was not to the state of nature, but to the meticulously subordinated cogs of a machine, not to the primal social contract, but to permanent coercions, not to fundamental rights, but to indefinitely progressive forms of training."[26]

There is an organic connection, Ursula Frankin concludes, between *designs for compliance* and a *culture of compliance*.[27] This sort of culture has deep institutional roots, particularly in an institutional town like Kingston, with its prisons, military base, hospitals, and asylum. The prisons were bastions of efficiency and Puritan notions of progress, with the daily activities of the "Inhabitants of Kingston" (as the convicts were delicately known) timed and regulated with ruthless precision.

Not long after Kingston Penitentiary was built, there were complaints about female inmates being flogged. The warden replied that the women were rarely beaten with a rawhide whip but were usually just thrown into a dark cellar and fed bread and water. The government responded to criticisms by writing down detailed regulations to systematize such matters. Some apostles of progress objected to harsh prison conditions and also preached against the demon rum. Historian Clare Pentland notes "a widespread campaign to render the labouring classes more industrious and obedient by removing liquor from their reach."[28]

In Kingston the modernized culture of compliance is reflected in an increasingly tenuous attachment to *place*. Universities and military bases attract people who are notoriously transient. The industries once firmly rooted in Kingston (locomotive building, shipyards) are gone, and manufacturers like Alcan, K-D Manufacturing, and Northern Telecom (which seemed rooted) are on the wane. Kingston's service industries, both public and private, reflect wider trends. This is particularly true of tourism, one of the biggest growth sectors of the late twentieth century. But tourism is scarcely a solid foundation for a secure future. It depends on the whims of visitors, and the work it offers is seasonal or irregular and usually not well paid. But most often there are few other options; people have little choice. If geographical location has become less important in the global scheme of things, what place do people have in the new world order?

Near the top of Kingston's north end is Rideau Heights, its ghet-toized concentrations of public housing mixed in with moderately priced subdivisions. This is where many of Kingston's poor are con-centrated; where, according to the lore of the respectable citizenry, the denizens are likely to be connected with the "Inhabitants of Kingston"—the pen, that is.

Not so, says Gary Dunn, who works in the social casualty ward's Intensive Care Unit. A community outreach worker with the North Kingston Community Development Project, Dunn grew up in the north end and got a job as a prison guard at age twenty, lasting three years before the pressure of the job took its toll. He reckons that about 30 per cent of those he meets in his work are transient or have some relationship with the prisons; they are either on parole or have moved to town to be near an inmate. The majority are simply the working poor and the welfare poor.

According to Dunn, by the time some of the kids hit fourteen, mom has lost track of them and has little idea of where they are or what they are doing. There are a lot of drugs around: coke, pot, hash, speed, some heroin. It is part of his job as a streetworker to warn north-end youth away from drugs. "Drugs aren't the answer," he tells them.

"Oh yeah, asshole. I make $800 a week [dealing]. What do you make? You think I'm gonna work at McDonald's with my hair cut short and look like an asshole?"

Like many people who do this kind of work, Dunn has recently been seeing more people who are on welfare for the first time. "They don't know where to start," he says. "I've had middle-class people in their thirties and forties sitting here crying."

This is the one side of the story of despair in the face of shifting employment prospects. According to Dunn, Kingston has a big problem with youth unemployment, with the welfare-beats-mini-mum-wage syndrome widespread. On the other side of the coin are the tough kids who don't cry, but are angry. But it is the false bravado of the young who, encountering the state in the form of the welfare worker, respond with an arrogance that masks their powerlessness:

"My fuckin' cheque better be ready or her ass is gonna get kicked." Gary Dunn's understanding stems from his everyday experience in the world of the underclass. It is a blunt message: "The youth don't have any hope."

Kingston is also a place where an old-fashioned Toryism contends not with the social democracy of a working-class movement but with a social liberalism often characteristic of college towns. Military and college reunions are more frequent than labour demonstrations; controversies over threats to historic buildings are likely to overshadow labour disputes. At the same time there is a relatively strong feminist movement. Helen Cooper, an active preservationist who came out of the university, served two terms as mayor before being appointed to chair the Ontario Municipal Board by the NDP government. Although local labour leaders bridle at the suggestion, she agrees that the unions have little power in the town. Cooper, who is not unsympathetic to unions, describes their influence as minimal. "Labour is not a voice in this community. It simply isn't." Kingston has always lacked the sort of powerful and self-conscious trade union presence that has tended to give an industrial city like Windsor a rough-and-ready egalitarianism. Apparently, for many years Kingston businessmen regularly noted "the reasonableness of the labour movement."[29]

The largest private sector employer in Kingston is Du Pont, a U.S. chemical colossus that cultivates a paternalistic image. Du Pont promotes company unions that do little to challenge its authority. "We're not company dominated," an official of one such Du Pont union once said. "We're so weak the company doesn't have to dominate us." In the late 1960s, two years after the United Mine Workers struck its Kingston plant, Du Pont managed to replace the UMW with a company union, the Kingston Independent Nylon Workers.[30]

These days Kingston may be serving as a laboratory for broader challenges confronting trade unionism, the social movement best positioned to protect the interests of people with "collars of all colours." But here, as elsewhere, traditional (male, industrial) blue-collar jobs in the sectors where trade unionism has always been strongest have been disappearing. The city's service economy, with its emphasis on tourism and public sector work, and its hopes of attracting new jobs in finance, computer services, and biotechnology, points in the direction of a labour market—and a society at large—dominated by small, hard-to-organize workplaces. Four out of five private-sector workers have no link to a labour movement that is scrambling to stop the erosion of its base.

In 1988 two long-time Kingstonians (one a geographer, the other a historian) published a book on Kingston called *Building on the Past*. They entitled their introduction "The Personality of Place" and observed that, rather than dubbing the town The Limestone City, its character could best be captured by a single word *continuity*. "Few cities have changed as little as Kingston," they said. The town has "a continued sense of community and a townscape of diverse domestic and institutional architecture . . . replete with open spaces bequeathed by past military and government functions and never far away from surrounding countryside and lakeshore vistas. It is an ethos to be cherished."[31]

Such a comforting description does indeed capture something of the place, an aspect that has attracted retirees and new silicon-based businesses, tourists, and artists. But this aspect clashes—as sharply as ever—with another reality. Another local professor perhaps put it best: "Workers have come to face hazards and stresses and setbacks that could hardly have even been imagined by those who struggled for working-class betterment before them. The present is not only different from the past, it is also almost always more complex."[32]

8

FLEXIWORKERS AND THE FUTURE

"All That Is Solid Melts into Air"

> If manure is suffered to lie in idle heaps, it breeds stink and ver-
> min. If properly diffused, it vivifies and fertilizes. The same is
> true of capital and knowledge. A monopoly of either breeds filth
> and abomination. A proper diffusion of them fills a country with
> joy and abundance.
>
> — *Poor Man's Guardian*, 1834

BY 1994 Canada's politicians were in bad odour, a pack of venal rogues intent on driving the country to ruin. Disrespect would have been a kindly description for what the citizenry thought of its elected leadership. A generally suspicious press sniffed the winds, eagerly reflecting the national mood back to the people.

With one exception: Frank McKenna, a political enterpriser with a keen eye for image, managed to sell himself as the salvation of his province. New Brunswick was open for business, selling itself to any company that cared to listen, attracting jobs, and trimming the fat to make it all viable. Investors could call 1-800-McKenna. The premier was so up-to-date that he had his own E-mail. He cleaned up the education system, courted private-sector employers, and talked regularly on a cellular phone.

In a typical bit of slack-jawed hagiography, *Maclean's* featured a cover story on "Fast Frank," presenting him as more or less a prototype of the yuppie go-getter extraordinaire. McKenna worked sixteen-hour days, he "breakfasted with officials from a high-tech company." He called civil servants at two in the morning. He lived in an unpretentious house. He cut thirteen hundred jobs from the civil service.

"Work becomes so obsessive that it dominates every aspect of your life," the premier said, his neo-Calvinist spirit bubbling over. "I am obsessed with creating jobs. I believe that it deals with the physical and spiritual needs of people." Fast Frank had "turned his province into Canada's social laboratory." *Maclean's* showed him running along in his cordovan loafers, clutching his important briefcase, wasting no time at "setting the pace for Canada."[1]

The pace would include the fast track to social reform. When Human Resources Minister Lloyd Axworthy announced a 1994 review of Canada's social programs—which he hoped would be the new Chrétien government's biggest project—he made explicit reference to New Brunswick's experiments with social security reform. In 1992 New Brunswick had embarked on a six-year, $177-million training project, with two out of every three dollars coming from Axworthy's department. The New Brunswick blueprint for linking social and labour-market policy, with its upbeat title *NB Works: Transitions to Self-Sufficiency by the Year 2000*, is larded with the new vocabulary of concern and participation. "As the industrial economy goes through rationalization and renewal," the project proposal begins, "averting . . . long term dependency on social assistance is a critical necessity to the future social and economic security of all New Brunswickers." In "the new economy" citizens would be "educated and adaptable," and "life-long learning" would be the norm. According to the proposal, "Many of the people dependent upon income support programs are people for whom the traditional approaches to education have failed and 'life-long learning' is a remote concept. As the requirements for educational attainment increase for permanent labour force participation, the likelihood of 'life-long dependence' increases for this population."[2]

The goals of this industrial strategy—and it is industrial strategy as much as social policy—are explicit. They are practical, ideological, and financial. First of all, the government wants to "develop the human resource potential" of welfare recipients "to achieve the goal of a more educated, better trained work force." Then there is the need to "change the attitude" of the poor that is "an end in itself" to an "attitude that people, though unemployed, must increase their employ-ability and job ready status." The final goal is to save social assistance costs by taking people off "the caseload" and getting them into work.[3]

The NB Works project targets those in need of basic literacy, a laudable goal. But its high cost—$59,000 per trainee—means that the project is unlikely to be duplicated in an era of spending cuts. What *will* be duplicated in one form or another is its emphasis on forcing people into the bad jobs end of a segmented labour market. The "incentive" will be either relatively high-priced carrots like intensive training or (more likely) the stick of punitive new social assistance regulations.[4]

The whole approach ignores a crucial reality: that the training will at best only prepare working-class people for "non-standard" or bad jobs, for marginal participation in a contingent labour force of just-in-time workers. It will prepare them ideologically to have diminished expectations. Training under this kind of social policy is industrial strategy by omission: the approach does not link skills acquisitions to the preponderance of bad jobs. The problem it addresses has nothing to do with creating good jobs, or with providing the unemployed or underemployed with decent jobs. Rather, it is about preparing people to be "job-ready" for whatever is available. It is about holding them close in the labour market's grey area between bad jobs and welfare. The goal, implicit or explicit, stated or understated, is to develop Galbraith's "functional underclass," ensuring that the people who inhabit the murky world of poverty or near-poverty are also "flexible."

Let us step back and consider, for instance, the difference in language and values between NB Works and the report of the Croll Committee on poverty. In 1971 this special Senate inquiry also talked of "dependency" and of Canadians being "trapped" in a nasty cycle of poverty. It also agreed that the social safety net of the time—which was at a high point from the perspective of the unemployed—was dysfunctional, "a chaotic accumulation of good intentions gone out of joint."

But David Croll and his colleagues did not share what is clearly one of the principal assumptions of those who would reform the welfare state of the 1990s, making it more "modern" and "efficient." The assumption has been around for a long time.

> Many cherished myths that helped give birth to the welfare system must be given final burial. One of these, that the poor are always with us, is a notion that the Committee categorically rejects. The economic system in which most Canadians prosper is the same system that creates poverty. Equally fallacious is the belief that economic growth could, in time, 'solve' poverty. The evidence produced before the Committee showed that in the 1950s and 1960s [when Canada enjoyed great economic expansion], in absolute terms poverty in Canada increased at the same time and at a similar rate.[5]

The essential point was that capitalism, even in the boom years, was socially unsustainable. Over two decades later Canadians have experienced major shifts in how capitalism operates. We have heard that global capital is increasingly footloose, with "no fixed address." Unemployment has risen. Some family incomes have kept pace, mainly because of the continuing shift of women into the workforce. But in spite of this the middle is being hollowed out of society. Two major downturns in the business cycle have produced deep recessions. Food banks are a way of life. The market is working its transformative magic.

In recent times no one represents the spirit of free-market capitalism better than Britain's Margaret Thatcher. In her autobiographical defence of her statement that there is "no such thing as society . . . there are individual men and women and there are families," Lady Thatcher pointed out that she had high regard for her nineteenth-century soul mates. "The Victorians," she explained, "had a way of talking which summed up what we were now rediscovering—they distinguished between the 'deserving' and the 'undeserving' poor."[6]

This is very much the venerable poor-laws-and-workhouses approach. For Thatcher and many of today's social reformers, it is okay to handcuff a large group of people to the lower echelons of the labour market, forcing them into low-paid jobs. Pundits call this variety of conditional approach "workfare." It is unfashionable in some circles because of its coercive connotations. The more liberal-sounding variants are also conditional, with "learnfare" benefits tied to enrolment in training. The most coercive approaches have certainly not disappeared from the political and policy landscape. In 1993 a new approach was introduced in Alberta, completing the circle: "busfare." Premier Ralph Klein argued that the best way to deal with the poor was to give them a ticket and send them packing to British Columbia. But the coercive aspects of learnfare are usually leavened out by an emphasis on participation and self-respect.

Such is the approach of the New Brunswick reforms that have in turn inspired Ottawa. According to McKenna's moral entrepreneuri-alism, social programs "make it comfortable for people to do nothing and learn nothing."[7] The social reforms of the 1990s tend to differentiate between the employable and unemployable poor, and in this they are a throwback to the past. Two New Brunswick critics of the McKenna reforms scolded the government for its attitude:

> By emphasizing the difference between unemployable and employable [welfare] recipients, you are reviving the notion that there are two groups of poor people—the deserving and the non-deserving. Such labelling often engenders harsh and punitive attitudes. . . . The employable (non-deserving) group is often blamed for the poverty it is experiencing in spite of the reality of

a sick economy that cannot accommodate them. The Poor Laws taught us that this categorization leads to one's "employability status" being the main criterion for assistance rather than "need.". . . People do not choose poverty and income assistance as a career goal.[8]

Robert Mullaly and Joan Weinman recalled that in 1960 New Brunswick had been the last province to abandon archaic Poor Laws, adding that many thousands of people working at poverty-level wages do not leave their jobs, and that many more unemployed search diligently for work.

"We are marginalized, have no control over flooded labour market conditions, and don't have access to employment alternatives," wrote an incensed Peter Marten, himself a social assistance recipient. Marten was particularly angry about the life-skills classes so popular with training advocates. He called the classes "welfare department sandbox and toilet-training boondoggles."

"We know how to budget, apply for job openings, tell time, tie our shoes, shop for food, cook it and even eat it too," wrote Marten in response to a McKenna welfare-reform initiative. He accused the premier of "abuse of power and lying."[9]

But so strong are the winds behind the training imperative that even the budget-slashing Thatcher government did not attempt to tack against them. It increased training spending in the mid-1980s, although, characteristically, it also privatized the training. In 1991 the Tories handed over control of their $4-billion annual training budget to employer-dominated Training and Enterprise Councils (TECs). In an approach advocated for Canada by the C.D. Howe Institute (whose director Tom Kierans is a huge fan of Frank McKenna), the British Tories instituted a voucher system through which young people could use the vouchers to shop around and buy training on the open market. The businessmen running the TECs had every incentive to "cream off the most easily-trained jobless." All of this prompted *The Economist*, a magazine normally sympathetic to the market's invisible hand, to observe, "Training is on the verge of collapse. Vulnerable groups such as the handicapped and ex-offenders will be the losers in the new training market."[10]

With the arrival of Chrétien in Ottawa and Clinton in Washington, the learnfare approach to social policy and the training solution to broader labour market issues gained more momentum. Both new governments appointed men well-known as liberals to look after the supply side of the labour market. At the same time they put more conservative ministers in charge of the treasury, where the really important economic policy decisions get made.

Clinton's labour secretary Robert Reich had underlined how unskilled and semi-skilled workers were "globally vulnerable" because they had to compete with huge numbers of people with the same skills in other lands. Reich pointed to trends being duplicated in other advanced capitalist countries, where workers "find themselves in an increasingly precarious position" with their "incomes slipping and their jobs disappearing." Reich knew that "extra training" could maybe slow down the loss of production jobs or the decline in real wages, but, he said, "It is far from a solution to the problems faced by unskilled and semiskilled American workers competing worldwide." His central insight was that the United States was becoming increasingly dependent on the well-paid work of his version of Lee Dyer's "P&Ts"—professional and technical employees—a group Reich labelled "symbolic analysts," the "fortunate fifth" of the populace who were growing ever more distant socially from less fortunate Americans.[11]

In Canada a savvy Jean Chrétien appointed Lloyd Axworthy as head of the new Human Resources Development superministry. Although not an academic like Reich, Axworthy was a smart, experienced politician, part of the faded leftish rump of a Liberal Party that had embraced continentalism (and by extension globalism) in the days of C.D. Howe. For a party that traditionally campaigned from the left and governed from the right, Axworthy was the perfect person to further erode the welfare state and legitimize the changes by showing a concern for the unemployed and underemployed, by sounding like a new-age dependency theorist.

Within a year after taking office (during which time the government made its first round of cuts to benefits for the unemployed), Axworthy's Human Resources Development ministry released a green paper on social security reform. Unemployment insurance would be changed to emphasize "individual responsibility and self-sufficiency." The green paper criticized the former welfare state as being too passive and promised a more active, "modernized" approach. The document recognized that both older workers and young families have been "squeezed out of the middle class" and that society was becoming more "polarized between highly-educated, highly-skilled Canadians in demand by employers—today's economic elite—and less educated people without specialized, up-to-date job skills, who have been losing ground." It was unclear as to whether the elite was composed of employers or the well educated, or both. In any case the whole social policy review generated much public interest, particularly in the context of the ongoing dispute between those who felt that the deficit was the enemy and those who argued that unemployment was the real problem. The issue of work, its distribution, and its design did not figure large in the debate.[12]

Axworthy's new ministry had been cobbled together in the dying days of the Mulroney government. HRD included welfare, labour, training, unemployment insurance—almost anything to do with the welfare state and the supply side of the job market. But the Liberal interpretation of social reform was not much changed since the Tories told a 1992 OECD meeting in Paris that they intended to emphasize "breaking the spiral of dependency" through "self-sufficiency." Axworthy himself would soon be referring to "the cycle of dependency."[13] Under both the moribund Tories and the new-look Liberals, there was to be a great emphasis on training.

Few would deny the importance of literacy and a good general education for all, particularly those ill served by a class-skewed public education system. After all, most poor people have poor educations. (As the 1994 green paper pointed out, poor kids perform worse at school than do children from wealthier backgrounds; some seven *million* Canadians have very limited literacy skills or difficulty with everyday reading.)[14]

But at a 1994 meeting on labour market polarization, sociologist John Myles described education and training as a *necessary* but not *sufficient* condition for personal prosperity. He gave an example of young women he has met selling clothes in Ottawa stores. He knows they are relatively skilled because they are university graduates—his former students. It is the jobs that they are doing that aren't skilled.

This point is crucial. We usually think of people as educated, trained, skilled. Jobs are, well, jobs. But just as people can change their skills, so can the skill content of jobs change; or, to be more precise, the skill content—and the wage paid—can be changed by employers. When most available jobs require little skill, all the training in the world won't help. The problem with "adjustment policy" is that too frequently it is people—their expectations, skills, feelings— who get adjusted. The way jobs and work are structured and distributed remains unchanged; the existence and further creation of so many bad jobs become an unalterable given.

The work that is available is increasingly poorly distributed, a fact that Myles and his colleagues at Statistics Canada discovered when they examined the shifting distribution of working hours in the 1980s. They concluded that it is "quite clear" that unequal access to *hours* of work was a major cause of increasing inequality of *earnings* in Canada; and that the use of a "core" of full-time workers, many of them working longer hours, "supplemented with increasing numbers of part-time, part-year or contract workers," provides companies with greater flexibility in their operations. They are also able "to focus training and advancement on the core employees."[15]

In 1984 two professors at the Massachusetts Institute of Technology published a weather-vane book that fascinated academics theorists and management types who spend their time trying to understand trends in production and corporate strategy. Interest in *The Second Industrial Divide* was a sign that people were searching for insights into how mass production of long runs of standardized goods (so-called "Fordism" after Henry Ford's mass production of Tin Lizzies) was giving way to what some called "Benettonism." This new "ism" took its name from the Italian garment company that is able to adjust production runs quickly, using subcontractors rather than its own workers to produce clothes in small batches, keeping inventories low, and using its computer-based market antennae to detect—and mould—shifts in consumer habits.

Michael Piore and Charles Sabel, authors of *The Second Industrial Divide*, argue that the old days of Fordism had been replaced by "flexible specialization," which they say is "a strategy of permanent innovation, accommodation to ceaseless change, rather than an effort to control it."[16]

Between 1979—when Sony introduced its Walkman—and 1992 the company introduced 227 different models of a product that had never existed before. Beer drinkers know all about this. As recently as the late 1970s they had the choice of a limited selection of brews that came in one standard bottle, the short, brown stubby. Within fifteen years Canadians would come to face a dizzying choice of lite and dry and ice beers, premium brands and small independent labels. Dozens of containers replaced the stubby.

Computerization provides a huge spur to flexible specialization, because equipment can be put to new uses producing new products without expensive and time-consuming rebuilding or retooling operations. This can be done by reprogramming. CNC (computer numerical control), CAD/CAM (computer-aided design and computer-aided manufacturing) and a host of other acronyms have revolutionized how employees in manufacturing do their jobs and how their bosses organize the work. Entire FMSs (flexible manufacturing systems) combine computerized machine tools, self-propelled trollies without drivers that supply parts to machines, and centralized computer controls for the whole operation.

But microtechnology, Piore and Sabel note, is not enough to drive the move to flexible specialization. "If the computer appears to be the cause of industrial flexibility," they conclude, "this is probably less because of its application than because, malleable as it is, it has helped crystallize the vision of a flexible economy just as the costs of rigidity were becoming obvious. Computer technology is a kind of magic mirror."[17]

If it is a magic mirror, it is one reflecting age-old trends and conflicts. Other observers of computers and of how the machines have altered the organization of work agree that those changes are not entirely due to computerization, but they are less sanguine about how computers have been deployed. Harley Shaiken, a former machinist who studied computer automation, points out that flexible manufacturing systems combine technical and social goals. "The drive to reduce direct labour is pervasive," Shaiken says. He quotes the vice-president of one firm that designs the systems as stating that "the flexible manufacturing system is really more of a management tool than it is a manufacturing tool." *Iron Age*, a trade magazine for steelmakers, puts it more bluntly: "Workers and their unions have too much say in manufacturing's destiny, many metalworking executives feel, and large, sophisticated FMS's can help wrest some of that control away from labor and put it back in the hands of management, where they feel it belongs."[18]

The word "flexibility" is very much like the word "progress." It can mean a lot of different things to a lot of different people. Both words sound positive. Progress is almost always assumed to be good. Similarly, flexible people and organizations are surely to be envied. After all, the opposites sound unattractive: rigid, sclerotic, unbending. But it all depends on where you are coming from—on whether you are viewing things from below or from above. Tom Peters, one of the most successful management gurus of the age, represents the view from above. Peters hit it big with the 1982 book *In Search of Excellence*, selling over five million copies. When many of his most excellent companies ultimately failed, Peters decided that there were no excellent companies and published several more books revising his theories. In 1992 he can · out with *Liberation Management: Necessary Disorganization for the Nanosecond Nineties.*

Nanoseconds are the unthinkably short units of time that computers use to make their calculations. Peters argues that everything is happening so fast these days that only flexible, lean, decentralized organizations can hope to cope with the need to change and adapt quickly. With his catchy subtitles suggesting disorganization ("Hi Ho Destructive Competition!" "Toward Productive Anarchy") and quotes from important Japanese design executives ("We do not seek permanence"), Peters presents a worldview shared by hordes of bosses.

What they *are* doing in the "nanosecond nineties" is trying to get rid of as many functions—and people—as possible. Such organizational flattening helps to explain the pervasive trend to contracting and subcontracting. Written in a manic style ("Hey! . . . Yikes! . . . Oops!"), the oxymoronic-sounding *Liberation Management* is uncharacteristically sober on the subject of firms contracting: "Assume for a heartbeat that it makes sense to consider subcontracting *everything*. Over the past couple of years we've challenged participants at our executive seminars to consider doing just that." However, Peters quickly reverts to form: "The cry from the executive suite should become 'Prove it CAN'T be subcontracted!'"[19]

The trend Peters spotted was becoming pervasive in the 1980s. It involves dismantling organizations, making them vertical instead of horizontal, although they remain hierarchical. Corporations have been dividing people up into small groups ("Teams . . . and More Teams") so they can compete not only with other businesses but also with each other ("Everyone a Businessperson"). The goal is to reproduce market relations internally, cutting jobs and fostering competitive anxiety everywhere.

This is a management universe in which flexibility is supposed to be king and speed is everything. "In 1987," Peters says, "I proposed *flexibility* as the watchword for the '90s. I'll give myself a nanosecond pat on the back. It was a start."[20] But when Peters and the rest stumbled upon the great new trends of flexibility, globalization, and impermanence, they were simply rediscovering long-established tendencies. Geographers Scott Lash and John Urry start their 1987 book *The End of Organized Capitalism* with a quote from a book originally published in 1848. "All fixed, fast-frozen relations are swept away," Karl Marx and Friedrich Engels wrote about the bourgeois epoch.

"All new-formed ones become antiquated before they can ossify. All that is solid melts into air, all that is holy is profaned."[21]

The quote has long been a favourite of radical interpreters of capitalism's historical mission and of how market ideologists ("Hi Ho Destructive Competition!") glorify the ways in which the innovations of capitalism reweave and so often destroy the fabric of people's lives. Whether it is the enclosure of common land or the introduction of Flexible Manufacturing Systems to displace people and skill, capitalism is revolutionary. But, again, revolutionary change always has at least two sides, the view from above and the view from below.

As capitalism evolved, and as what became known as the industrial revolution took shape, voices from below questioned progress. They questioned how their lives were being reshaped, their worlds turned upside down. The Luddites of early nineteenth-century England have never enjoyed a good reputation: mindless machine-smashers who supposedly tried, stubbornly, pathetically, and ultimately hopelessly, to stand in the way of progress. Empowered by new machines, factory owners were destroying the skills, jobs, and working lives of hand-loom weavers and other craft workers, replacing them with the sweated labour of factory women and children. Those who resisted, often by destroying machines, were doing the only thing they could to try to stop the apparently irresistible force. Anyone caught going to a meeting to form unions could be arrested and transported to Australian penal colonies, or worse—Luddism often ended on the scaffold. Yet David Ricardo, one of capitalism's earliest theoreticians, declared simply that opposition to the new machinery was "conformable to the correct principles of political economy."[22]

In the 1980s, when it was becoming clear that structural joblessness was becoming a permanent feature of Canada's social landscape, neither business nor government had the inclination to do much about the demand side of the labour market. As a result there was one overwhelming answer. Training advisory boards, training task forces, training commissions, and committees combed the offices, industrial plants, and colleges of the land in search of the perfect "learning corporation." Canada needed new skills for its "new economy."

"I even had one manager tell me they weren't a factory anymore, they were a university," recalled David Robertson, a researcher for the autoworkers union.[23] The CAW was watching its ranks being decimated by closures in the auto-parts sector, downsizing in the transportation sector, and the shutdown of the east-coast fishery.

On the other side of the bargaining table, the chairman of General Electric's Canadian subsidiary emphasized his firm's commitment to both training and "empowerment" of its workers. Gone are the days of business using "large numbers of low-skilled employees," said Dennis Williams. "The low-skilled, well-paid work of [the] post-war era has been designed out, automated or transferred to low wage countries." According to Williams, whose firm eliminated the jobs of a hundred thousand people in the 1980s, it is crucial to have multi-skilled workers who work at processes that are "as simple as possible." He says, "Complexity is often a by-product of insecurity. People who were unsure of themselves or their worth in our organization tended to surround themselves with complexity."[24] He was perhaps referring to some of the hundred thousand no longer with his organization.

Some companies are frank about their training goals. Northern Telecom, for instance, produces complicated digitalized switching equipment, a field in which the company has achieved worldwide success. These machines need to be carefully tested by experienced troubleshooters who iron out the bugs before the product is shipped. This can be complex work but, like generations of bosses before them, Northern management has tried to simplify it as much as possible, perhaps echoing the spirit of the Toyota approach to training: "We just try to keep it simple."

According to Northern workers, testing finished circuit boards was once a complicated job. Workers made autonomous decisions. Then the company broke the work down into separate operations. "Now you just have to push a few buttons and that's it. The test set does it," one worker complains. "It tells you where it's bad if it's bad, so you hand it to the repair operator. . . . They've deskilled the job."

This is where the F-word comes into play. "We want to reduce tests to their lowest common denominator," a candid Northern manager admits. "It increases our control and our flexibility with labour. It's much easier to find lower skilled people."[25]

What does this mean for the new high-skill, high-wage economy that is one of the advertised results of global competition for those willing to train and train again? At the Northern plant in London, Ontario, employment declined from a peak of 2,300 in the 1970s to 700 in 1993. The remaining jobs were sharply divided between complex tasks and repetitive jobs that require little skill. Telephone sets were easier to manufacture, with standardized subassemblies that arrived "just-in-time," often from overseas. The London workforce adapted to the changes. Some workers found themselves sandwiched between two robots on an assembly-line doing repetitive, routine tasks. Others relearned their jobs. Those who had taken years to learn to set up a machine to produce a part for a telephone had to keep up by adapting to new keyboard-based automated systems. Even though the members of Northern's skilled, high-tech workforce in London did everything they could to adapt to change, it served them little good. In the end the company decided to close the operation and relocate production elsewhere.

We often regard brain work as challenging and interesting, but workers who spend their days monitoring electronic symbols are dealing with abstractions. They have no control over what goes on behind the computer screen. That is the preserve of far-off systems designers. Much work—like that at Northern—is no longer as physical or tangible as it once was. Such work may be called "hi-tech."*

---

* Indeed, the telecommunications work relocated to New Brunswick has often been described as "high-tech" in spite of the fact that the key factor in the relocation was the province's high unemployment/low-wage climate (plus government subsidies of $6 million). When the UPS courier company moved its call centre to New Brunswick in 1995 it was not unaware that its competitor Purolator was paying between $8 and $9 per hour wages in that province.

But that does not mean it is by definition challenging or interesting: it means the operator has to be in a constant state of alertness while monitoring the screen, ready to follow a predetermined set of instructions if a particular message appears. There is little or no participation in planning or designing the work.

According to a union study of work at three Northern plants, including the cable factory in Kingston: "In automated workplaces the change in jobs is not exclusively in one direction, either for better or for worse. Instead there is a polarization of effects. The changes associated with new technology at Northern have led to skilled and complex automated work and to unskilled, repetitive and fragmented jobs."[26] From the company viewpoint, this means increased "flexibility" with labour. Such flexibility, internal to the firm, reflects the broader good jobs/bad jobs segmentation of the workforce at large. This inequality has not only grown in measurable ways, including the shrinkage in the number of jobs that pay something close to middle-income wages. But also central to the rise of "disorganized capitalism" is a change in the nature of employment relations.

When a firm such as Northern "downsizes" (by going global and shifting work abroad or by contracting-out work that used to be done on an in-house basis), it is changing the way it operates: moving from a system of wage employment in which protected, full-time work is negotiated between centralized firms and large unions to a system characterized by less collective and more individualized arrangements. The growth of contractualized employment erodes the system of *collective* bargaining so dear to the hearts of trade unionists. There is a far less secure form of relationship between those who buy and those who sell labour. In the new world of work, individual people who want to sell their services do so on a casual, contract, subcontract, part-time, or temporary basis. People in the expanding pool of flexiworkers ofttimes perform a combination of these, so it is possible for an individual worker to be on one or more part-time, temporary contracts. This complex brew also includes the rise of the informal or grey economy untouched by government regulation (taxation, UI payments, pension plans).

This trend—to individualized and insecure work arrangements—is what Guy Standing of the International Labor Organization calls "subordinated flexibility," underlining how the new flexibility is very much on the employers' terms. His terminology suggests a hierarchy that simple *flexibility*, a positive-sounding term, often masks. What is happening is more than income inequality; it is inequality of social class. Flexibility is often about power—about who is losing it and to whom.[27]

Jobs and bargaining power have most obviously been lost by what Standing describes as "the old proletarian stratum," that group of mainly male workers who in Canada drew their strength from unionized collective bargaining for good wages and benefits in return for operating jackleg drills in the mines, bucking timber in the forests, or bolting bumpers onto cars on the assembly-lines of central Canada. The conventional wisdom is that the days of "Fordism," of high wages for what is called unskilled work (typified by Henry Ford's assembly-line) are over as jobs become automated, with capital replacing labour in the mines, forests, and factories. The forest industry, historically Canada's most important source of jobs and wealth, is a case in point. As the forest frontier is pushed back and more land is logged, fewer jobs are produced. Employment in the pulp and paper sector is predicted to decline from 72,000 in 1991 to 52,000 by the end of the decade.[28]

Such workers have not disappeared, nor have their unions been fatally weakened. As labour-intensive production is automated and shipped offshore, a highly qualified core workforce will remain, along with many less skilled workers. Unions will also remain, although their members will face constant pressure to be "functionally flexible." But the union capability of maintaining decent wages will not be matched by the ability to protect job security or resist management demands for "multi-tasking" that further erode on-the-job power.[29] Off the job—in the community at large—unions risk becoming isolated from the anxious army of workers, particularly young people and racial minorities, trapped in a low-wage/no-benefit cycle of despair. This latter group may resent what they see as the privileges of organized workers. The segmentation of the labour market may give rise to divisive cleavages within a working class that already reflects the diversity of the population at large.

Consider the images of the darkened factories of something called the "rust belt." Now you can tour a modern paper mill without seeing any workers on the plant floor, because the survivors of automation stare at computer screens monitoring programs written and controlled by Robert Reich's "symbolic analysts." It is now practically impossible to scan the ads on public transit vehicles without seeing private-sector come-ons for temp agencies and training academies. Virtually every union in the land has come to focus its bargaining goals on job security.

If the old proletarians represent a shrinking core of a dual labour market, the flexiworkers of the new service proletariat are multiplying. Many training programs of the 1990s are designed to equip people with the attitudes and aptitudes to fit in as part of this stratum. It is here that employment contracts are individualized, small business dominates, unions are weak or non-existent, and many women work. As large private-sector and public-sector employers move to contracting-out, more people find themselves competing for work as contract cleaners, security guards, and clerical workers. Jobs scatter about like so many shards of glass from a bottle that has crashed to a cement floor. This fragmentation forces labour to scurry around to grab up the little bits of work at the bottom of the labour market.

Much training—like that at CDI in Kingston or Windsor's Unemployed Help Centre—is aimed at the flexiworkers of the service proletariat. This training seeks to impart (limited) computer aptitude in which people learn particular programs. But just as important is the ideological training—the sort of attitude adjustment that instructs people to be ready for the chase and not to expect anything more than the chase. Attitude adjustment in the guise of training is largely unrecognized, but it is also offered to workers still employed in the good jobs sector. General Motors calls it "cultural training."

Northern Telecom speaks of developing within its (remaining) flexible workers "a strong corporate culture" that "encourages the workforce to accept the company's goals as its own."[30] A Windsor worker who was shunted from a full-time job in manufacturing to part-time work in a nursing home puts it another way: "Being flexible means always staying limber and being able to bend over. Because the drive will always be from behind."

———————————

The symbol of youth employment as the millennium approaches is the burger-flipper. It describes those consigned to the famous "McJob," the antithesis of a solid job.

The image of a burger-flipper can bring to mind a short-order cook working behind the counter at an old-style diner, that remarkable grill magician capable of juggling a dozen different orders, preparing eggs-over-easy-bacon-well-done and tuna-salad-on-brown-hold-the-mayo while harried waitresses shout still more orders.

The reality of the McJob is, however, quite different. In fact, even the term "burger-flipper" is inaccurate, at least as it applies to those who toil at Burger King's broiler-steamer work station, where workers feed pre-measured frozen patties onto a conveyor belt leading to an automatic broiler. Manufactured by a Burger King subsidiary, the broiling machine takes exactly ninety-four seconds to turn the puck-like disk into a cooked burger. The unit is capable of regurgitating 835 cooked patties from its "meat chute" every hour. At the end of the line a worker uses tongs to place the burger on the heel of a bun. Burger King bosses usually put the newest, least experienced workers on the broiler-steamer because the jobs are so simple and easy to learn.

Firms like Burger King are not terribly concerned about workers quitting such mind-numbing jobs because the workers are easily replaceable. The companies are, however, anxious about customer loyalty. Although the fast-food chains don't have to compete for workers, they do have to fight with each other for customers. "Why Customers Quit" is a crew-room sign that states in no uncertain terms that most burger buyers "quit" because of "attitude indifference" on the part of workers. The corporation trains the staff to conceive of burger-buyers as "guests."

Burger King workers learn that they are part of one crew that pulls together. During busy periods, managers roam the floor saying "Let's go, team." The Burger King message to its workers is: "Your job is sort of a social occasion. You meet people—you want these people to like you, to like visiting your restaurant."

This notion of workers as *teams* does not mesh smoothly with the thinking promoted in the training courses offered to the unemployed: the approach of rugged individualism and the war-of-each-against-all. The notion of teamwork is also not confined to the burger-flipper jobs at the bottom of the service sector. "Teams" are supposedly operating at all levels of the job ladder. Human Resources managers interested in Total Quality Management promote Quality of Working Life programs as they strive to convince their underlings that the workplace is like a football field. Managers are coaches who "stay on the sidelines and send in the play."[31] Factories become universities, and fast-food joints are teamwork nirvanas.

A silky vocabulary of empowerment and ownership (". . . visiting your restaurant") pervades the human resources field. The ideal workplace for the year 2000 is being imagined as a co-operative spot where equal parties will exchange suggestions for improving quality. Many managers, influenced by the high-priced consultants who roam the industrial relations field like snake-oil salesmen in the Old West, adopt "mission statements" and talk of self-awareness. Even the most lean bureaucracies, public and private, seem to find the resources to retain the high-priced services of consultants. (Incidentally, these consultancies, as "business services," are often seen as one of the most dynamic growth areas in the economy.) The human potential set, it seems, has somehow infiltrated the world of Human Resources.

For instance, in one of the most famous cases, Shell Canada hired consultants to set up an experiment in workplace reorganization at its chemical plant in Sarnia, Ontario, and the main message these consultants delivered was that both sides should "forget union and forget management and work together."³² According to *Toronto Star* writer David Crane, "Anyone who wants to see the future of work" should begin by looking at the Shell plant, a model for good jobs prosperity in a high-wage, high value added economy. "Employees largely run the plant" and help to train one another. The union (a local of the Communications, Energy, and Paperworkers Union, or CEP) has a say in hiring. There are no more supervisors, just "co-ordinators" who work as management reps inside the work teams. Gone are reserved parking places for managers and management dining rooms. (Gone as well in this model are many managers; if you can get workers to supervise each other in teams, you can do some "delayering," a word that means the same thing to managers that "downsizing" means to workers.) Computers provide each worker with regular, up-to-date information on what's happening, easing communication. The newly "flexible" workforce is a good alternative to the old "authoritarian structure of command."

In this world of newly empowered workers, Crane suggests, there is no real need for unions, although such partnerships "will have a much better chance of success if employees are able to exercise their collective voice through healthy labour unions."³³

Workplace democracy and worker control were once cherished goals of the worker's movement. During the days of the Paris Commune, the greatest urban uprising of the last century, the workers there experimented with taking complete control, prompting the anarchist Bakunin to dream of a society "organized from the bottom upwards."³⁴ Trade unions are based on the idea that the collective voice of organized workers is inevitably stronger than the feeble strength of one. A thorny question that divides Canadian union leaders is whether new, top-down ideas about teamwork are the wave of the future.

In a close look at both the Shell Sarnia experiment and other human resource schemes, Donald Wells showed that the worst fears of those unionists opposed to team concepts were realized at Sarnia after a few years. "It appears that the teams have replaced the union," concluded Wells, who teaches political science at McMaster University. The role of the shop steward, the crucial on-the-job union representative, was dramatically reduced, with the chief steward reporting that by the late 1980s the project had "started to drift badly." The company preserved a modicum of labour peace, avoiding layoffs of the core workforce by subcontracting maintenance work to outsiders who can be laid off when they are not needed. According to a local union leader at an Eldorado Nuclear plant in Port Hope, Ontario, the effectiveness of the local declined after the team program was introduced. "Back in '78 we were militant," he recalled. "Now you can't lead the members across the road."[35]

A similar scheme introduced at Xerox (the firm whose president denounced other firms for spending less on training than he does dining out) involved training in "interpersonal skills" and "how you look at your job." Such is the job insecurity in the business that the Amalgamated Clothing and Textile Workers (ACTW) agreed to a flexibility scheme under which Xerox brings in temporary workers who are paid five dollars an hour less than the regular workers and have no benefits. When the company laid off the temps, one union leader claimed that since they were not full-time people, "Nobody got hurt."[36]

Canadian unions appear to be divided—at least on a rhetorical level—about how to react to "lean production"—the generic term that became commonly used after the 1990 publication of an influential MIT study of sixty car plants around the world. *The Machine That Changed the World* was so enthusiastic about Japanese methods of organizing work that it actually recommended that "the whole world should adopt lean production, and as quickly as possible."

The physical machine in question is the automobile, but lean production is based on social machinery, on how people are organized to work. So it is really about technology in Ursula Franklin's sense of the word—*design for compliance*. It means being obliged to work faster and harder and, following the Japanese model, constantly cutting the employer's costs. If the goal of eternal speed-up seems questionable, lean production enthusiasts offer the promise of "endless product variety" as compensation.[37] This is a worldview in

which a proliferation of constantly changing commodities and freedom to choose among them have been elevated to the highest pinnacle of social value.

Call it lean production or call it Toyotism, but management has
adopted a new terminology for its old task of forcing or cajoling people to work harder and faster. Take the idea of "kaizen" or continuous improvement, a notion that has seeped its way into management
consciousness. A sure indication that it had caught on with management consultants came when the word began to be used as a verb (as
in "visioning" or "impacting"). A job is kaizened if it is made faster,
more efficient. Kaizen is said to "empower" workers, because bosses
hope to use their workers' insights to eliminate waste.

The showcase CAMI plant in southern Ontario produces a line of
cheap, tin-can jeeplets for its owners, General Motors and Suzuki. It
boasts work teams and kaizen, the whole gamut of Japanese production techniques. Workers trained to do not just one job but a range
of repetitive tasks rotate through a series of simplified, easily learned
procedures. Part of kaizen training involves convincing the workers
to let the boss in on ways in which the work can be made even faster
and simpler, thus eliminating waste—wasted steps, wasted arm
movements, and so on. When this happens, their team size is
reduced and the intensity of work increased. Sociologist James
Rhinehart, who participated in a multiyear study of workers' shifting
attitudes to lean production at CAMI, found that when kaizen
worked, workers worked harder. They also got to eat in the same
cafeterias as the managers.

"Workers that we interviewed indicated very clearly," Rhinehart
said, "that continuous skills development is not occurring, the access
to training is very limited, and that they're just rotating these highly
repetitive jobs."

One word that employers insist on using to describe workplace
life is "quality"—as in Total Quality Management or grouping workers into Quality Circles. But such quality-from-above has a narrow
meaning. Firms who brag about quality ("Quality is Job One!") are
essentially concerned with the production of commodities that are

free of defects, to avoid having a product come back to them during the short life of the product's warranty. Quality in this sense has little to do with products that are durable or needed.

———————

Inglis, an old name in the "white goods" or appliance business, is owned by a U.S. outfit called the Whirlpool Corporation. Whirlpool has closed several venerable Canadian plants. The year the free trade agreement was signed, it shut down a Toronto plant in which three generations of workers had been making washing machines, throwing 650 people out of work. The average age was forty-seven, the average length of service fifteen years. Production was relocated to Clyde, Ohio. Inglis then put up a billboard on the roof of the abandoned Toronto plant to take advantage of its location near the busy Gardiner Expressway: "KNOWLEDGE IS POWER."

Pat Kopita makes garbage compactors for Inglis in Cambridge, Ontario. She was forced to take a pay cut and chase her job to Cambridge when Inglis shut down its operation in Stony Creek. Her workplace is divided into a conventional assembly-line that produces clothes dryers and the newer, brightly coloured "Hi-Commitment" (HI-C, for short) line where she and her fellow workers put together garbage compactors. A compactor essentially replaces the human foot in households in which people realize the importance of keeping down the volume of waste they haul out to the curb. Later on after a few years' use, when the compactors inevitably break down and are not worth fixing, they will surely join the rest of the litter of consumerism in our overflowing garbage dumps.

On the HI-C line Pat Kopita stands on a small wooden platform covered with frayed Astroturf, adding prepainted top and side panels for the compactor. The workers still do repetitive tasks, but these are carried out by "work teams" with six or seven employees each. Team members rotate jobs, determining scheduling, work assignments, and co-ordination of their small section of the line. Cobbling together the compactors is like assembling the same Meccano set again and again. Workers screw metal parts together, insert the motor, add handles and knobs, and test to see if the thing works. The clanging of metal parts is punctuated by a screaming whir-and-screech as workers tighten bolts and screws with pneumatic tools. The sound is enough to make a dentist's drill seem soothing.

There are no "lead hands." There are no foremen. Above the work area there is, however, a flashing message board, just like the ones in bank windows that advertise the latest RRSP and mortgage rates. The message at the Inglis compactor work area blinks out the number of units the line has produced and the number of flawed machines. It also adds with monotonous regularity "Good Work!" Another sign proclaims that "Today's Quality is Tomorrow's Customer."

Kopita acknowledges that the HI-C line, which she describes as a "push-and-pull" operation because she pushes and pulls all day long, is less boring than the conventional assembly-line. A half-dozen workers make twenty-seven units a day, and they can see the finished product and test it. The workers are responsible for ironing out any production or organizational problems they encounter.[38] Kopita explains that the compactor line and the HI-C approach were part of a package that the corporation offered as part of the increasingly common technique of getting workers at different plants to bid against one another in a kind of race to the bottom. The Cambridge workers were told that if they didn't go into HI-C and the trash compactors, their jobs would head south to another Whirlpool plant in Danville, Kentucky. Having seen Inglis shut down in Toronto and Stony Creek, Kopita recalls the level of anxiety at the time. "We never had a choice," she says matter of factly. "We signed."

Ron Bureza, Training and Development Co-ordinator for Inglis in Cambridge, says that some workers in the unionized plant are used to the adversarial system of labour-management relations. People have mixed feeling about the new Japanese-style management techniques. He agrees that the widespread insecurity at Inglis about the future of jobs is paradoxical, making it both easier and harder to get people involved. Is it a double-edged sword, forcing people to grasp at straws and at the same time making them mistrustful? Bureza believes that despite all the talk about free trade, "Any company that can produce the best quality product at the lowest cost is going to be the one that's around." The focus should be on making a better quality product than the competition, at a lower cost. The plant's education should focus on that too.

"As a result of the insecurity people feel," Bureza says, "they're looking at education as a tool to enhance their skills so they can carry them with them when this place closes down. I can understand that. But I don't think that should be people's motivation. We should all be striving to make the top quality product at the lowest cost."

At Inglis employee involvement, like the jobs themselves, is narrowly defined. It does not extend to traditional management rights such as when, where, and how new technologies are introduced, the overall design of the workplace, the skill content of jobs, or even how the surplus generated by the Cambridge plant of the Inglis subsidiary of the Whirlpool Corporation (the world's largest maker of major appliances, plants in a dozen lands) will be distributed. Will it be used to provide basic literacy and skills training to Canadian workers or to pad the stock options of the executive board in Benton Harbor, Michigan?

A 1993 survey of 714 Canadian business establishments revealed that decisions on everything from capital investment to training programs and work teams were overwhelmingly concentrated in the hands of top managers. Training co-ordinators like Ron Bureza and workers like Pat Kopita are rarely if ever involved. The survey showed, not surprisingly, that 99.3 per cent of investment decisions were made by top management. But workers, unions, and even "human resource managers" were virtually shut out, even when it came to planning training programs or deciding whether to bring in work teams. Workers and unions had a say in training in exactly 0.4 per cent of the cases. When it came to work teams, the people on the teams were involved 2.1 per cent of the time. By contrast, top managers made training decisions in 87 per cent of the cases and team decisions in 93.8 per cent.[39]

According to an Inglis newsletter (*Committed to Excellence*), the firm has "core values" like teamwork and a "mission" to be the biggest supplier of white goods in Canada. It also wants to "empower our people to continually improve quality." And, of course: "We are part of the Whirlpool Corporation recognizing that . . . company strategies cross borders and are designed for the good of the whole."[40]

Pat Kopita was forced to join the experiment in Hi-Commitment teamwork under threat of losing her job to the United States. As a member of a "lean team," when she misses a shift because of illness or union business, it puts a lot of extra pressure on her "teammates." She no longer has a supervisor on her back about this. The teams are allegedly autonomous, with management reluctant to spring for replacements. So Kopita's new supervisors are her own co-workers. "When I'm away, it makes it harder for them," she says. "Instead of the team working with six people, it was working with five. They started complaining and yelling at *me*."*

Peter Larson, an expert in improving business-labour dialogue, explained that the Inglis scheme "dramatically altered work relationships . . . between employees and the boss." In 1995 the boss decided to end that particular relationship. Whirlpool closed its Cambridge plant, throwing Pat Kopita out of a job as part of a worldwide restructuring that eliminated thirty-two hundred jobs.[41]

Employers are undertaking the mishmash of training, work reorganization, and lean production schemes at a time when their position has been strengthened by the conjuncture of high unemployment,

---

* Management professor and consultant Lee Dyer is sanguine about the nature of managing people by grouping them into teams. In his 1993 Donald Woods Lecture in Kingston he stated clearly that teams are *not* "a soft way to manage," adding pungently that "those of us who have had teenage children know what the notion of peer pressure is all about."

globalization, recessions, and the general revolution of falling expectations. These conditions present an enormous challenge to the union movement.

But while Canadian unions are back on their heels, they are not totally off balance—particularly in comparison to their U.S. counterparts. Many (and particularly the strongest and most self-confident), reject both narrow, instrumentalist training as well as the notion that they should act as one with management-sponsored teams in a fight against workers around the corner or around the world.

At Inglis the HI-C program gave rise to a major internal dispute over how to react. Apparently, both workers and their various unions have different responses to lean production. Few talk tougher than the Canadian Auto Workers union. The CAW rejects any hint of a team scheme. "We were one of the first unions to adopt a national policy statement critical of workplace changes which came under the heading of team concept," it announced in 1993.[42] The union was a powerful force at CAMI, touted as a bridgehead for Japanese-style co-operation. CAMI did not have workers, just "production associates" trained for teamwork. But when a strike hit in 1992, *The Windsor Star* reported on a clash of attitude: "Young, idealistic, bright—they began at the CAMI Automotive plant believing this was no ordinary factory. It's three and one half years later, and those 2100 workers are now cynical and embittered, wielding picket signs that ridicule the workplace philosophies they once embraced."[43]

The CAW won a remarkable victory in that strike at CAMI, where forty-three thousand people had applied for just over two thousand jobs in 1987-88.[44] But a CAW local at Honeywell was forced to live with a TQM program. In 1993 the plant employed 470 production workers, who turned out as many thermostats as 890 CAW members had done just five years earlier.

"The enemy is outside," said Lloyd Miller, president of the CAW Honeywell local. "The people who are trying to build a better product than you are and take your job away, that's the enemy."[45] Faced with a multinational employer that can maintain output while eliminating half of its workers, these unionists have been forced to play ball with the company on what is clearly not a level playing field, even though two CAW staff intellectuals have written, "Once we decide to play on the terrain of competitiveness, we cannot then step back without paying a serious price."[46]

There appear to be differences between this position and that of the U.S.-based Steelworkers union, which has also had both successes and losses in dealing with work reorganization. A 1992 Steelworker guide to work reorganization sounds more accommodating than the 1993 CAW statement, which says that lean production proposals need to be "resisted or changed."[47] The Steelworkers seem more open, stating that many locals want to take advantage of workplace change and "develop positive proposals that will change the way work is organized, the way our members do their jobs, and enhance the role of the union."[48]

According to the Steelworkers, a local might want to work with management to reorganize production "to build a stronger union, a stronger company, and better jobs." The Autoworkers would not be caught dead making formal statements about building stronger companies, but in fact they do say much the same thing using a different vocabulary. Although they don't want to be "junior partners in production," they do want to build stronger relations with the boss to improve "the productive capacity of the workplace" and gain better working conditions.[49] The similarities may be more important than the differences. Both unions realize that negotiation of Japanese-style changes depends—like any other negotiation—on the relative strength of the players.

In the end, what people hear on the street and in the pub, what they see on TV, what's happening to the expectations of their kids—these will likely do more to determine the reactions of workers to new employer schemes than the policies of their unions—if indeed they have unions.

The very term "lean production" brings to mind more stale, but nonetheless sobering, phrases: trimming the fat, downsizing, delayering, belt-tightening; in other words, tossing people onto the street, or reducing them to part-time, temporary status. "In a context of overcapacity, international competition, high unemployment, declining rates of unionization," labour-watcher Donald Wells said, "much of the stability of lean production depends on labour weakness *outside the labour process.*" Job fears are perhaps the most crucial element in this weakness. Getting workers on the job to pressure each other to work faster has little traction unless it is accompanied by fear of unemployment.[50]

Unions that know they have a strong hand to play can afford to participate in the risky game of work reorganization. Big industrial unions like the CAW, Steelworkers, and CEP *may* be able to confirm the cliché and turn co-operative programs into a "win-win" situation. But even more than in traditional collective bargaining over wages and benefits, making gains—or just surviving as a strong voice for the membership—depends on how well rooted the workers' organization is among the actual workers. When bosses offer to "empower" their employees, they are not talking about giving them a collective power. Despite all the talk about teamwork, this form of empowerment bears an individualistic stamp, steeped in the lore of competition, pitting worker against fellow worker. This means that for unions and individual workers it is a risky game to play, with many potential losers.

Weaker unions like the ACTW or the massive public-sector unions whose members often work in scattered and isolated settings may be more vulnerable to the predations of management-sponsored team proposals. In such cases the alphabet soup of TQMs can result in shifting attitudes and the replacement of the union team by the employer's team. The key is the involvement, participation, and control of the union by its members. If a union as a collective organization is deeply rooted in the working lives and the consciousness of its members, chances are that it will remain strong in the face of the most sophisticated work reorganization plans. Some unions have also realized that they have to counter management-training courses with education programs of their own. Commitment to broad-based membership participation and education is more than ever a litmus test for union strength.

However, with the decline in employment (460,000 Canadian jobs disappeared in 1989-93; unemployment in OECD countries rose from twenty-four to thirty-five million from 1990-94) and the rise of a good jobs/bad jobs labour market within Canada, the labour movement is in trouble. The growing segmentation of a split-level labour market means that fewer people are protected by a collective voice on the job.

Many of the good jobs that have disappeared have been union jobs. Private-sector unionization dropped from 25.7 per cent in 1975 to 20.7 per cent in 1985. John O'Grady, an industrial relations specialist and former union staffer, predicts that if trends continue unionization in the private sector could slide to 16 per cent by the turn of the century. The only thing standing between the Canadian unions and their devastated comrades south of the border is Canada's highly organized public service. Between 1975 and 1985 the proportion of union workers who held government jobs rose from 42.1 to 56.4 per cent.[51]

The unions that have not seen their numbers reduced have had to run faster just to stay in the same place, by organizing new workplaces to make up for the losses to technological change, free trade, contracting-out, and so on. For O'Grady, "A wedge is being driven between the unionized minority and the non-union majority." Collective bargaining does not have an even impact on the labour market. According to O'Grady, "For two-thirds of workers, collective bargaining affords no direct economic protection and increasingly less indirect protection." O'Grady worries that more and more non-union workers, with their "stagnant real wages," will come to blame the unionized workers for achieving gains at their expense.[52]

What's more, the rise of part-time and temporary work, contracting-out, and all the other hallmarks of a polarized, just-in-time workforce means that workers will be harder to organize than those in the big plants and hospitals of the past. Between 1978 and 1986 the proportion of employees in workplaces with less than twenty people rose from 16 to 24 per cent.[53] Some of these small establishments are indeed small businesses, but it would be a mistake to see this trend

being driven by a nation of shopkeepers, mom and pop plant nurs-
eries, and one-truck movers. Many small workplaces, from branches
of the Royal Bank to Burger King, are controlled by the most powerful multinationals around. Forming unions in such places is difficult
and expensive, as is servicing them if they are successfully organized.

Canada's declining rate of what specialists in the area call "union
density" is most often compared to the situation in the United
States, where unions are in much sharper decline, where the differences between the rich and the poor are so wide and growing so fast
that political observers like Kevin Phillips have devoted entire books
to the dire implications of a declining middle, a society polarized by
fear and the politics of frustration as it teeters on the edge of the next
century's silicon-chip economy.[54] These tendencies are mirrored in
Canada. The politics of frustration with declining living standards,
the constant pressure to sell yourself in a competitive job market
(often populated with people who look different and speak different
languages), a tax system skewed towards the wealthy, and a generalized anxiety about the future: all of these elements have provided fertile ground for the nostrums of a Reform Party that combines social
conservatism (send women back to the home and more people to
jail) with economic liberalism (get government out of the way and
let the market decide). This is the context for the Americanization
(or erosion) of the welfare state, the emphasis on an individualistic,
survival-of-the-fittest approach to social policy.

It is hardly surprising that Canadian unions most always compare
their fortunes with those of their U.S. counterparts. But it might be
more fruitful to look to Europe, where (as in Canada) the union
movement is not only stronger but is also open to more creative
responses to the crisis. German trade unions were estimating in the
1980s that business and government had thus far employed only a
small fraction (about 5 per cent) of the new technologies that would
be available by the year 2000. The reserves of productivity in both
manufacturing and services are hardly imaginable.

Wolfgang Lecher of the German union central DGB's Institute of
Economic and Social Research agrees with John O'Grady: "The

opposition between labour and capital is increasingly coming to be overlaid by an antagonism between workers in permanent, well-protected jobs on the one hand and peripheral workers and the unemployed on the other. . . . The trade unions run the risk of degenerating into a sort of mutual insurance for the relatively restricted and privileged group of permanent workers."[55]

As part of the great postwar accommodation (the "deal" described in chapter 1), unions gave up their historical challenge to the rights of management and their long struggle for shorter hours in return for job stability in a high-employment, high-consumption economy. The prosperity that was to be the basis for the deal was based on the traditional staples economy as rocks, logs, and oil continued to flow from the hinterland while the branch-plants of central Canada supplied the domestic market. By the 1960s, with the expansion of the welfare state, the unionized workers in the mines, forests, and factories were joined by legions more from offices, schools, and hospitals. In 1947 less than 30 per cent of non-agricultural workers were organized. By the mid-1970s the total had reached 37 per cent, and the Canadian Union of Public Employees was the largest in the land.[56] Since that time, things have changed, with a steady rise in unemployment and a levelling off of union growth. By the 1990s the deal was off.

For unions, and the rest of us, the time has come for a shift in thinking. If productivity is to skyrocket and employment is to plummet, does it really make sense to push for "full employment" and the freedom *to work*? Rather, the need for freedom *from work* as it has traditionally been defined becomes more apparent each time we hear that poverty and economic growth are simultaneously on the rise. But when they hear complaints about joblessness, the powers that be haul out predictable figures showing the *number* of jobs being created. Rarely is there any mention of the *kinds* of jobs. Still more rare are discussions of what all this new work is accomplishing or contributing to the common good.

"If critics of the prevailing orthodoxy continue to put full employment on the highest pedestal of social policy," Guy Standing observed, "they should not bellyache if regulations are adapted to promote that goal."[57] By the mid-1990s regulations were indeed being adapted in Canada. The regulations, dressed up by politicians' speechwriters in terms of self-reliance and of turning the social safety net into a springboard, were in fact aimed at handcuffing the growing numbers of poor people to growing numbers of low-wage, dead-end jobs.

With the postwar deal cancelled, some people began to think again about returning to what the workers' movement had, for its part, given up. What about the rights of management? How about the struggle for shorter hours, for freedom from work? Perhaps, as Andre Gorz puts it, "To reject work is also to reject the traditional strategy and organizational forms of the working-class movement. It is no longer a question of winning power as a worker, but of winning the power no longer to function as a worker."[58]

Historian Clare Pentland has noted the importance of the changes in the "nature of the supply of, and demand for, labour" in influencing how economic restructuring—something we hear much about at the end of this century—took place in the nineteenth century. The changes also brought "a transformation of values and attitudes." Analysing the shift from the staples economy to the industrial economy that characterized Canada's industrial revolution, Pentland describes independent loggers who had not yet been subjected to the discipline of the clock—to say nothing of the computer that measures time in nanoseconds.

The loggers "were efficient, not from a devotion to a religion of capitalism but as a matter of personal pride. They resisted retrogression, but they did not know that man must progress. Their choice of occupation was determined primarily by local custom." They also worked for bosses who "understood the pre-industrial values of fellowship, prowess and tradition." According to Pentland: "The lumber boss who could break a log jam, knock a man down and lead a song, could expect more enthusiastic production than the flabby competitor whose range of interest was from cover to cover of the account book. But the future belonged to the latter."[59]

The movement towards a culture of compliance—the tick of the clock or the clang of the factory bell—was bound up with changes in both the *expectations* of working people and the hierarchy of control over their work. Today's euphemism for the account book is "the bottom line." We are constantly reminded of how it drives every cost-cutting and staff-reducing decision. Just as the rise of the production manager who worked in an office far from the forest frontier signalled "the transformation of Canadians," the rise of the global corporation that deploys robots and shifts its capital with the speed of light is also having its effects on the nation's citizens. For legions of Canadian job seekers, experience has taught them about what German writer Hans Magnus Enzensberger describes as "that little difference between first class and steerage, between the bridge and the engine room."[60] For many Canadians, that "little difference" makes all the difference in the world.

9

WORK, TIME, AND THE WHEEL OF FORTUNE

The machine [has] penetrated everywhere, thrusting aside with its gigantic arm the feeble efforts of handicraft. . . . After a century and a half of labor-saving machinery, we work about as hard as ever. For the great majority of the workers, the interest of work as such is gone. It is a task done consciously for a wage, one eye upon the clock.

– Stephen Leacock,
The Unsolved Riddle of Social Justice, 1920

THERE WAS A TIME (and it was not really that long ago) when workers and assorted radicals and socialists would protest the gap between poverty and wealth by railing against "the idle rich." This class of indolent coupon-clippers, *rentiers*, and aristocratic spongers had little moral traction; if they bothered to respond to critics who denounced them for getting so much for doing very little, the idle rich could only invoke some sort of "natural order of things."

In this scheme of things workers were in a position to occupy the

high ground. They could point out, in the words of the old union
hymn, that:

> It is we who plowed the prairies, built the cities where they
> trade . . .
> All the world that's owned by idle drones is ours and ours
> alone . . .
> They have taken untold millions that they never toiled to
> earn . . .
> But the union makes us strong! Solidarity forever . . .[1]

Things have changed. Nowadays we hear about the punishing
schedules of the well-to-do. The lawyers and managers and consultants, the professional and technical people ("P&TS"), Robert Reich's
"symbolic analysts," such people work sixty, seventy—even 80-hour
weeks. They have an arduous workload, but they know how to get
things done, too. The rich apparently deserve their good fortune.

In a travel article aimed at upscale couples, journalist Judith Timson offers poignant testimony to this, outlining "energizing" quickie
vacations. She and her husband had tried it, and it worked. Her
advice to fast-track careerists was to be goal-oriented in their leisure.
"Stay focused" on shopping (one recommended destination is Min-
nesota's 360-store Mall of America—the "ultimate shopping binge"),
sightseeing, eating, whatever. Jammed between ads for vacations in
the Cayman Islands and India, the piece is essentially a come-on for
medium to high-priced tourism (with a bill of up to $1,200 per per-
son for two or three days of relaxation).

We learn that "life in the nineties means more work and no play,"
especially for the self-employed and the tense survivors of corporate
downsizings. "We're in hot pursuit of the one commodity eluding us
all these days—not money, not even happiness, but time."[2]

What this amounts to is the remoralization of the rich. Not that
hard-working urban professionals see themselves as "rich" in the old
liveried-chauffeur sense of the word. Harried representatives of the
upper reaches of the middle class doubtlessly regard themselves as

part of a comfortable class of people deserving of everything they have. After all, they work so hard for their money that they have little time to spend it. From this perspective, it is not hard to imagine that the people at the other end of the social scale—those whose *surplus* of time may take, say, the form of underemployment in a part-time job—comprise the undeserving poor.

The issue of access to work is important in any consideration of a shrinking middle of the workforce. In the United States, differences in wage rates have been responsible for much of the erosion of middle-level income. But in Canada access to working time has been "a major determinant of the growth of earnings inequality."[3]

For Dave Lachapelle, time has never been a commodity. Back when he worked at a monster press stamping automobile bumpers out of sheet steel, he could never understand the men who grabbed every hour of overtime they could, working Saturdays and Sundays whenever possible. He detested overtime "with a passion." One of his favourite clauses in the union contract at Windsor Bumper was the one stating that overtime was strictly voluntary.

Lachapelle, a soft-spoken, reflective man, tells about the only time he went all-out for hours at work. He had angered some of his fellow workers with his outspoken attacks on long hours and figured that if he spoke out about an issue he should have first-hand experience. So for six months he worked seven days a week, twelve-hour shifts, double shifts back to back.

Was I right in what I was saying? I found myself drinking a little more than I would normally do. I found myself snapping at my family, which wasn't my personality. I found myself looking at the calendar to find out what *day* it was. I worked holidays—triple time! What more incentive do you need? But I reached a point—the second time I almost fell asleep at the wheel on the way home—that I pulled over, and to be perfectly honest with you I cried. I was just so physically and emotionally exhausted. It scared the shit out of me that I had almost fallen asleep. I sat there and cried like a baby. I asked myself, "Am I losing my fucking mind?" I went home, gathered my thoughts, spoke to my wife. And everything I had been saying about working long hours was right. There is no question.

Still, Windsor Bumper never had a shortage of volunteers for overtime when it wanted to keep production going. Sometimes the men would sneak cases of beer into the plant. One fabled tradesman supposedly had a mickey of rye hidden in every electrical box in the plant. But the employees didn't stick around for eleven-hour shifts on weekends because they were having a good time or liked the mind-numbing work. Some of them felt they just had to work long hours to maintain a certain lifestyle. For others the job had become all there was.

Lachapelle had worked over half his life at the bumper plant when corporate restructuring and free trade closed it down in 1990. Now he has a part-time job in a nursing home within walking distance of the mobile-home park where he lives with his family. He has mixed feelings about the change.

"I've lived it," he says, recalling the frantic, lucrative days when the company was going flat out to fill a big order that would allow Ford to fit another line of Fairlanes with heavy, chrome-plated bumpers that shone like silver. "I know why it was that when I used to say that I had spent a nice weekend with my wife they'd look at me and say, 'What the hell you want to stay home with the old lady for?'" Still, he misses his old job, with its regular paycheque and sense of fellowship. He says that going to work is important: "You create a camaraderie, a society. It's like going to the local bar, it's comfortable."

Lachapelle was among others who occupied the plant in 1981 when the company threatened to close the plant if the workers didn't make contract concessions. Although he had no love for the work, he was attached to the others at the plant and the security of a regular job. "To have that torn from you is . . . is . . ." Lachapelle hesitates as he tries to describe the sense of loss. "Well, it's hard to explain what you feel."

The workplace that was such a large part of his life got flipped around so often during the casino capitalism of the 1980s that when he went to work he often didn't know who the new boss actually was. The former autoworker is not too pleased that the company he sustained for a quarter-century is still making bumpers down at the other end of US96, in Grand Rapids, Michigan.

At the same time, Lachapelle never did buy into the work-to-spend cycle of the goods society. His father put in thirty-four years at Chrysler, finishing up with a good job in the stock room, away from the assembly-line. His son lives at home, taking some upgrading courses while earning a few bucks at a local pizzeria. "Society says to you: 'If you've made it, then you have the two cars and you have the boat and not the hundred-thousand-dollar house, because that's average, but the quarter-million-dollar house.'"

Lachapelle, a conscious anti-consumer, laughs at the irony of it. "You're programmed to strive for that. Seven days a week and you have the means to get the whole cake. All of a sudden I'm eating angel food cake with the most wonderful icing. And then someone says, 'Well you can have the cake but a smaller piece. And forget the icing.'" For Lachapelle and others, the abrupt shift is hard to take. "Once you've had something and it's gone," he says, "that's worse than never having had it at all."

As the great recession of the early 1990s gave way to a painfully slow jobless recovery, financial analysts shrugged about the lack of "consumer confidence" and the apparent reluctance of the citizenry to do their bit by getting out there to spend. The recession had been the most severe since the Depression of the 1930s, whose end Ford celebrated at the 1939 New York World's Fair with a Cycle of Production exhibit that showed automobile manufacturing from mining to assembly. In this great cycle workers were nowhere represented.

By 1964 another New York Fair had rolled around; Dave Lachapelle was getting ready to drop out of high school for that secure future in a car plant. At the fair General Motors' corporate fantasy saw a future in which a jungle road-builder would take care of the rainforest. Historian Michael Smith has probed this form of corporate futurism:

> GM's vision of taming the jungle focused on replacing its natural transportation medium, an "aimless wandering river," with modern superhighways. First, a jungle harvester felled great swaths of trees with laser beams. Then the area would be sprayed with chemical defoliants, and "a road-building vehicle as high as a five-story building and as long as three football fields" leveled the cleared ground, set steel pilings, and extruded a multilane highway "in one continuous operation!" GM press releases predicted that this massive "road-builder," powered by its own mobile nuclear reactor, would be "capable of producing from within itself one mile of four-lane, elevated superhighway every hour."[4]

By the 1990s, when something called the "information superhighway" had replaced the road through the rainforest as the latest pathway to prosperity, the psychic and environmental pain associated with conventional views of progress was becoming apparent. More Canadians had begun taking seriously Dave Lachapelle's ideas about the relative importance of work. Periods of high unemployment often breed discussions that question our need to pass so much time at work, whether we are stamping out bumpers or (at the other, more rarefied, end of the split-level society) spending frazzled, sixty-hour weeks and then making a fast-paced getaway to the Mall of America or "high tea at low tide" in Bermuda.

When high and intractable levels of joblessness are combined with overwhelming evidence of environmental destruction and a cultural crisis that combines rising crime and the uncertainty about both private family life and public civility, people begin to search for answers. Some rally around the flag of "family values" and around putting more people in jail for longer periods, while letting the growth of the free market sort out the tiresome problems of joblessness, poverty, and pollution.

But will faster growth ease the pressure on stressed-out families, the forests, or the ozone layer? Will it ease the pressure that pushes people to scramble for jobs in jails in a land that already spends over $7 billion annually on prisons and policing?

In his book *Working Harder Isn't Working* British Columbia employment counsellor Bruce O'Hara outlined a "Too Much Trap" characterized by three tendencies: 1) overproduction, which causes unemployment to rise; 2) unemployment, which allows employers to drive wages down; and 3) workers who must buy less even though those still working produce more and more.[5] O'Hara's plea for a wholesale rethinking of the world of work was part of an ever-louder chorus that has linked joblessness with environmental and cultural issues. Most borrow (either implicitly or explicitly) from the ideas of thinkers, such as Ivan Illich, who question the values—family and otherwise—of growth-addicted consumer capitalism. O'Hara came up with one of the longest subtitles in Canadian publishing history: *How We Can Save the Environment, the Economy and Our Sanity by Working Less and Enjoying Life More.*

Economists such as Juliet Schor no longer risk being excommunicated from the polite circles of academic society when they point out that the treadmill of life in fin-de-siècle North America traps a majority of people into overwork while a growing minority languishes in the enforced idleness of unemployment. Schor's book on the unexpected decline of leisure, *The Overworked American*, points out what many, particularly working women with children, already know. Modern life is lived too quickly; too many people suffer from a shortage of time, even though it should be possible to produce 1948's standard of living in less than half the time it took that year.[6] Schor, however, is not mesmerized by images of happy info-workers puttering away in electronic cottages, plugged into the information highway. The twentieth century has witnessed profound shifts in the role of women's labour, which have lately been accompanied by changes in the labour market as a whole.

Women's time became an artificially undervalued resource. In exactly the same way that we use up too much clean air and water because it has no price, the housewife's time was squandered. . . . Ultimately, inequality of time must be solved by readdressing the underlying inequality of income. Only when the poorest make a living wage can their right to free time be realized. And barring an economic miracle, part of it will have to come from the people at the top. In the 1980s, the rich grabbed a fantastic amount from those below them. Now it's time to give it back.[7]

The issue of work time cannot be separated from the vision and values of consumer culture. Any movement towards less work will require not only work-sharing, but also wealth-sharing. Otherwise the gap between the haves and the have-nots in the alleged new economy, between those who work a lot and those who work too little, will only continue to widen. While the future may not be Brazil's high walls topped with broken glass, Canadians can anticipate L.A.-law, with personal safety privatized as those who can afford it huddle inside gated, guarded suburbs.

Many of us remember learning that the crucial invention of the industrial revolution was the steam engine. We are now told that the computer is the key to the second industrial revolution. Indeed, being trained to use a computer is held up as a prerequisite to success in today's intensely competitive labour market. In a comparison of the two inventions, British design engineer and trade unionist Mike Cooley pointed out that the steam-driven engine was working for 102 years after James Watt built it. The computer he was using in 1984 was obsolete in three or four years.[8]

According to Cooley, the winner of the Alternative Nobel Prize for his designs of socially useful products, people once had skills, and tools to put those skills to use, that lasted a lifetime. When today's most important tools—the ones that drive the second industrial revolution—become obsolete faster and faster, "so too do the skills that people require to use them. They are trained to use a particular piece

216

of equipment, but that knowledge is only valid for about two or three years."9

There is a crucial element at play here: speed and acceleration. Time has long been used as a tool of social analysis, because everyone experiences it; yet it is also an abstraction. Is time real or imagined? Can it be spent or wasted? One thing is clear: discussions of time bring out conflicts between basic values.

One of the chief characteristics of the changing labour market has been the rise in part-time and temporary work. Just-in-time production is said to be the hallmark of efficient, flexible organizations. The ability of currency speculators and insurance-claim processors to move information instantly across the world is a hallmark of a globalized world in which money traders can quickly drive down the value of a nation's currency and the medical claims of Ohio steelworkers are processed in County Cork, Ireland.

One of the most famous essays on the subject was written by Lewis Mumford in 1934. In "The Monastery and the Clock" the U.S. cultural critic and historian of technology pointed out, "The clock, not the steam-engine, is the key-machine of the modern industrial age."10 As Mumford noted, centuries before James Watt's engine the first mechanical clocks regulated the daily routines in medieval monasteries. Today, when steam locomotives are consigned to museums and the older among us can only nostalgically recall the haunting call of the steam whistle, the clock is still the predominant machine in our homes. Middle-class North American homes may each have three colour televisions that may or may not bring them into the 500-channel universe. Perhaps they are looking forward to home shopping along the information superhighway. But it is a virtual certainty that clocks—bedside digitals, blinking readouts on the VCR, wall and stove-top clocks in the kitchen, internal nanosecond regulators in the home computer—outnumber any other bit of technology in the home. Despite anxiety about the corrosive cultural effects of television, the enormous social and cultural implications of strict timekeeping are unsurpassed—and too often unscrutinized.

Mumford understood this well. For him, the coming of the mechanical clock—first to the monastery and then to the bourgeois town—meant that lives once played out to the natural rhythms of the season and the harvest were changed forever: "The clouds that could paralyze the sundial, the freezing that could stop the water-clock on a winter night, were no longer obstacles to time-keeping: summer or winter, day or night, one was aware of the measured clank of the clock. The instrument presently spread outside the monastery; and the regular striking of bells brought a new regularity into the life of the workman and the merchant."[11]

Social struggles over the new commodity have persisted ever since. Who would control the hours, minutes, and seconds that were the products of the new technology? Was this new regularity a good thing? By the fourteenth century, mechanical clocks were in regular use in Europe.

At the end of that century the dominant system of labour time was still controlled (as it would remain for many generations in many lands) by agrarian rhythms "free of haste, careless of exactitude," in the words of historian Jacques Le Goff. Land was divided and named a *journal* according to the amount that could be ploughed in one day, *un jour*. The way people worked mirrored the society of the day, "*sober and modest*, without enormous appetites."[12]

This way of doing things was under assault. The cloth trades experienced some of the first upheavals over time and its use; and it is consistent with today's ambivalent attitudes to overtime hours that, like so many of Dave Lachapelle's workmates, some workers wanted the working day to be lengthened so their incomes would rise. (On this point Le Goff cites the fullers' assistants in the famous French tapestry town of Arras.)

But opposition to the discipline of the town's *Werkglocke* (work-clock) persisted. The expanding cloth trades were the "leading edge" of the day, playing a role similar to today's information enterprises. The cloth-making bourgeoisie enthusiastically embraced the *werkglocke* as a means of controlling the work of their subordinates; time became a social category, "the time of the cloth makers."

"Worker uprisings were subsequently aimed at silencing the *Werkglocke*," Le Goff observes. The cloth-manufacturing bourgeoisie protected the work bells with zeal; the authorities did not hesitate to invoke the death penalty against anyone who called for revolt, not only against the king, but now also against the officer in charge of the work bell. "It is clear that in the late fourteenth and early fifteenth centuries, the duration of the working day rather than the salary itself was the stake in the workers' struggles."[13]

Five hundred years later, that prototypical revolutionist, the anarchist agent provocateur in Joseph Conrad's 1907 novel *The Secret Agent*, was instructed by his paymaster to blow up the Greenwich observatory, the most obvious symbol of timekeeping at the turn of the twentieth century: "It will alarm every selfishness of the class that should be impressed. They believe that in some mysterious way science is at the source of their material prosperity. . . . Yes, the blowing up of the first meridian is bound to raise a howl of execration."[14]

Today's computers and quartz timepieces measure time more accurately than scientists were able to do by observing the heavens through the telescopes at Greenwich. Not that such fractions of a second are much to quibble over, but the period between the medieval *werkglocke* and today's cheap quartz watch was fraught with conflict over time, as those who would *pass* it were ranged against those who would *spend* it like the currency described by Benjamin Franklin ("time is money"). A printer by trade, Franklin was used to customers who wanted their work done without delay. It soon came to pass that the symbolic recognition of decades of disciplined service to an employer took the form of a gold watch.

The principles of adherence to a timetable, originally enunciated in the monastery, were rigidly applied in schools, workshops, armies, and hospitals. Michel Foucault has pointed out that well before scientific managers brought a rigid division of labour to Ford's assembly-line, soldiers and students were subjected to the rhythms of signals, whistles, bells, and orders that "imposed on everyone temporal norms . . . intended both to accelerate the process of learning and to teach speed as a virtue."[15]

Teaching (or, perhaps, "training") people to internalize the discipline of the clock constituted a massive cultural change, a characteristic of industrialization that E.P. Thompson described as a "severe restructuring . . . a new human nature." The English historian showed that in the same way that Algerian peasants under French colonialism sometimes saw the clock as "the devil's mill," artisans in eighteenth-century England were reluctant to give up their precious Saint Monday, even under the shrill exhortations of various Methodists and merchants. Their work was irregular, and that was the way they liked it. Monday was "Sundayes brother" according to an old satirical rhyme, a day set aside by Sheffield cutlers and Yorkshire miners for leisure, rest, and personal business. It was not until the onslaught of the Victorian era, when, according to Dickens, "the deadly statistical clock . . . measured every second with a beat like a rap upon a coffin-lid," that the dominance of time discipline and the separation of work from the rest of life began to appear complete. Mill-owners would steal time from employees by setting the clocks forward in the morning and backwards in the evening.

Max Weber's description of the capitalist ethic drew on Ben Franklin's "time is money" dictum to describe its essence. According to Thompson, Franklin was a man of the New World, "a world which was to reach its apogee with Henry Ford."

Although we are told that we have now reached a post-Fordist world of flexible accumulation and globalization, we would do well to heed Thompson's thoughts on time, work, and life:

> If we are to have an enlarged leisure, in an automated future, the problem is not "how are men going to be able to *consume* all these additional time-units of leisure?" but "what will be the capacity for experience of the men who have this undirected time to live?" If we maintain a Puritan time-valuation, then it is a question of how this time is put to *use*, or how it is exploited by the leisure industries. But if the purposive notation of time-use becomes less compulsive, then men might have to re-learn some of the arts of living lost in the industrial revolution: how to fill

the interstices of their days with enriched, more leisurely, personal and social relations; how to break down once more the barriers between work and life.[16]

The lunacy of lives driven by the compulsions of work, speed, and consumption is evident everywhere, particularly in "the machine that changed the world." Automobiles are designed for maximum speeds that are not safe, legal, or sane. Other less obvious cultural manifestations lie behind the events that most fascinate us, from sports to disasters.

The decades-old fascination with the fate of the luxury liner Titanic has been the subject of books, films, articles, and underwater adventures aimed at photographing and even raising the wreck of the White Star liner whose owners ordered the captain to steam at full speed into a fog-shrouded ice-field in 1912. The ship's owners hoped that the largest moving structure ever made would be able to shave a few hours off the transatlantic crossing and thus attract more customers. The chilling image of the Titanic's huge steel stern as it rises hundreds of feet into the air before taking more than fifteen hundred people to their cruel, silent grave has haunted imaginations ever since that fateful night. But the words of the Chicago bishop who condemned the arrogance of the "insane desire" for speed have almost been forgotten.[17]*

Other odd desires persist. The fastest, most prestigious and most expensive way to fly the Atlantic is aboard the supersonic Concorde. This modern version of the Titanic, designed and guided

---

* Also more or less forgotten are the less-celebrated fates of the thirty Filipino sailors who perished in 1994 when the bulk carrier Marika 7 steamed out of Sept-Iles into the teeth of a vicious North Atlantic gale. The event rekindled concern over safety of crews on ships whose owners operate on schedules so tight that the on-time delivery of iron ore takes priority over human life. When Rev. David Craig, director of the Halifax Missions to Seamen, complained after the Marika 7 disaster that captains are routinely pressured to sail aging ships into violent storms so that owners can avoid cash penalties for late delivery, he was sacked from his job.

by sophisticated computer technologies, was produced by firms that clearly saw the potential demand for this sort of thing at a time when the same tools and talents could have been put to use making handier and cheaper—though slow-moving—wheelchairs. What priorities stimulate Concorde? More efficient converters could use the sun's energy to charge batteries for better wheelchairs for those millions of the globe's disabled who can't get around. Four of five Canadian families own a microwave oven.[18] What Canadian home, once accustomed to it, would easily give up the microwave oven and retreat just a few years to those days of cooking more slowly?

When viewed through an environmental or cultural lens, the compulsory consumption of superfluities makes little sense. For instance, a glossy magazine ad for a cellular phone portrays a designer toddler atop a playground dinosaur, his happy, GORE-TEX-clad mom steadying him with one hand while holding her Nokia mobile phone to her ear. It is not clear whether her smile is being stimulated by her child or the phone conversation. Equally (and, one guesses, intentionally) ambivalent is the headline chosen by the advertisers hired by the Finnish company to sell the Korean-made phones to Canadians: *Some things are just too important to miss.* Like "quality time" with the family? That important call? One thing that is undeniable is that there's no need to waste time looking up the number or dialling; the Nokia boasts alphanumeric memory and speed-dialling.

The accumulation of microwaves and cell-phones, like the accumulation of property and capital itself, has no apparent limit. There's a gnawing anxiety about the future of the family at the same time that the market burrows its way like a tapeworm into the guts of our everyday lives. The imperatives of compulsory consumption undeniably lead to families spending less and less time together. Less and less often do we share a meal. Rather, we prefer to "graze" by popping something hurriedly into the microwave before heading off to our meetings, TVs, and organized distractions.

"The family meal was once a primary family sacrament, where children learned the terms of civil discourse," Robert Bellah and his co-authors say in their compelling study of the social ills of modern life, *The Good Society*. "What happens to the family when commodification reaches this extent?"[19]

In his appeal for less work and more life, Bruce O'Hara suggests that the central symbol of our time is the refrigerator-door timetable, a chart that guides dual-income families through the shoals of life in the late twentieth century. Many parents adhere to rigid schedules for driving the kids to school, getting to work at one or more jobs, shopping for groceries, organizing music lessons, and so on. Marriages have eroded when people have been too distracted to notice. "Parents don't have time to be parents," O'Hara concludes, pointing out that smart parents slot in time together. "They try to make it up to their children with horse-riding lessons, Nintendo games, designer clothes and 'quality time.'"[20]

The social struggle over time has proceeded a long way since the time of Saint Monday, with many people coming out losers. According to Statistics Canada, one in three Canadians feels "constantly under stress" trying to do more than they can handle. One in four sees himself or herself as a workaholic. When presented with the statement "I often feel under stress when I don't have enough time," 45 per cent (and more women than men) agreed. Nearly as many (44 per cent) said that when they needed more time, they skipped sleep. One in five had resolved to slow down the following year.[21]

A profile of Canadian family life from the Vanier Institute of the Family shows that more than a third of dual-income families would sink below the poverty line if one partner stopped working. According to the research, family incomes are going "nowhere fast" and licensed child care is "not much, not cheap." Divorce is on the rise, and after separation women are much worse off than men. But along with these commonplace realities of the 1990s comes the Institute's description of a typical workday of a Canadian family.

> Up early to get the kids dressed ... breakfast eaten, lunches made, animals fed, kids delivered to daycare or school, and in to work on time . . . the commute home—stressful enough, even without

traffic jams or remembering to buy milk—pick up the little ones from daycare, prepare a reasonably nutritious meal while juggling phone calls, the latest mechanical calamity, and the children's problems. . . . If things go well, the kids will watch TV quietly so the parents can get the meal on the table as quickly as possible. [The Institute optimistically assumes that dad is helping with the cooking.] And then a leisurely evening at home? Hardly. Instead, it's baths, homework, a quick load of laundry because someone needs that special shirt the next day. Or maybe it's hockey, ballet or music lessons, or 4-H for the kids or a community college course in data processing or business administration for their parents to upgrade career prospects. And don't forget the parent-teacher meeting, or the community daycare meeting. And that exercise class to try and get the body in shape to keep up with this ridiculous pace! . . . It would seem that all too often today's families must live on the left-overs of human energy and time.[22]

The good life or the goods life? In his dissection of the language of cultural transformation, Welsh social critic Raymond Williams probed how the notion of "the consumer" had been moulded by the market. In its original English use, borrowed from the French at the time of the medieval *werkglocke*, "to consume" had negative implications. It meant to destroy, to exhaust, to use up, to waste—a sense that it retains today. In the middle ages, if a judge ruled you a heretic or a witch, you might well find yourself sentenced to being "consumed by fire." Tuberculosis was long known as "consumption." From the sixteenth century, when the noun "consumer" was used it had similarly wasteful implications.

It was only in the twentieth century that "the consumer" was transformed into an abstract figure in an abstract market. We now speak of "consumer-led" recovery and even have Consumers' Reports and a Consumers' Association. Williams situates the change in the word against the background of a particular stage of capitalist development and the needs of a mass market for artificially created needs. "It is appropriate in terms of the history of the word that criticism of a wasteful and 'throw-away' society was expressed somewhat later, by the description **consumer society**," he concludes. "To say *user* rather than **consumer** is still to express a relevant distinction."[23]

———————

A Catholic friend of mine tells about something that happened on one of his family's traditional Sunday drives in the country, something he has always remembered. As he peered out the car window he noticed a field littered with thousands of round stones. He asked about the meaning of this strange sight. "The farmer was working his field on Sunday, instead of taking a day of rest," was the response from the front seat. "All of the potatoes were turned into stones."

Ever since then, after years spent studying classical philosophy and working on issues of social justice, my friend says he has been puzzled about why something that seems so eminently reasonable—one day of rest out of seven—should need a commandment to back it up, or labour-standards legislation to enforce it.

As it is, we move faster and faster, only to remain in what Juliet Schor calls "capitalism's squirrel cage." Labour-saving devices, however many we consume, have apparently done little to save time. But Schor points out that consumerism is not an inescapable fact of human nature. The waste that has accompanied commodity culture has been with us since at least the 1920s, when productivity growth began to be translated not into relaxation and leisure for all, but into a culture of unlimited desires. According to Schor, "Business was explicit in its hostility to increases in free time, preferring consumption as the *alternative* to taking economic progress in the form of leisure."[24]

We are now living with the legacy of that choice. It takes the form of being afraid of the sun because the ozone layer is thinning and searching for new garbage dumps to replace the ones overflowing with ancient rusting gas barbecues and the packaging from the new home computer.

If business has always opposed reductions in work-time in favour of more work for more consumption, labour's relationship with working hours has been more ambivalent. Workers historically have wanted to work shorter hours. The response from above has been equally predictable. As early as 1816 Nova Scotia adopted a vicious anti-union statute that made unlawful meetings and association aimed at cutting hours or raising pay punishable by three months in jail. When Kingston workers agitated in the 1830s for shorter hours, the Loyalist press responded by blaming the situation on "calculating Yankees" who wanted to promote their "pernicious" ideas about "Atheism, Republicanism and Revolution." Fully a third of all strikes in the 1860s and 1870s were provoked by the desires of workers to get shorter hours or control over some other aspect of work life.[25]

The most important of these strikes took place in 1872 as part of an upsurge of labour agitation aimed at the nine-hour day. The movement had started among English construction workers in 1859, spreading across the Atlantic. A U.S. labour reformer of the day, George McNeill, explained, "Men who are compelled to sell their labour, very naturally desire to sell the smallest portion of their time for the largest possible price. They are merchants of their time. It is their only available capital." Throughout central Canada nine-hour leagues sprang up. At a Hamilton demonstration in 1872, workers led by railway machinist James Ryan used a horse-drawn wagon to display a gravestone bearing the epitaph "Died 15th of May, the ten hour system."[26]

Although the Hamilton workers and their allies were rebuffed in that attempt to work less, in the decades that followed unions pressed the demand relentlessly. In 1870 the standard work-week in Canadian manufacturing was sixty-four hours. But workers made major strides, particularly in the years immediately following World War I.[27] In the wake of the wave of postwar militancy, the eight-hour day became widespread.

"It may very well be that an eight-hour day will prove, presently if not immediately, to be more productive than one of ten," said Stephen Leacock as he gazed with alarm at the labour unrest. "But somewhere the limit is reached and gross production falls. The supply of things in general gets shorter. But note that this itself would not matter much, if somehow and in some way not yet found, the shortening of the production of goods cut out the luxuries and superfluities first."[28]

By the 1930s the work-week had declined to forty-nine hours. In a huge victory in 1937 the International Ladies' Garment Workers Union mobilized five thousand women to strike in Quebec. The ILGW succeeded in cutting the women's weekly hours from an astonishing eighty to forty-four.

A major watershed period was the 1920s, when labour demands intersected with a realization on the part of business that a consumerist consensus was vital to continued profitability. Advertising took off, along with instalment selling and consumer credit. At the same time that they learned about the dangers of "office hips" and "underarm offence," women suddenly found out that their proper place was in the home. It was there, within the family, according to home economist Christine Frederick (author of a 1929 book, *Selling Mrs. Consumer*), that women qualified for the rank of "quartermaster rather than general." Frederick's ideal woman runs the supply room "for the very reason that she can't lead the forces in the field."[29]

Also in 1929 a Committee on Recent Changes appointed by U.S. president Herbert Hoover summed it all up by proving "conclusively" that "wants are almost insatiable; that one want satisfied makes way for another." The committee concluded that economically the United States had "a boundless field" lying in wait in its future. "There are new wants that will make way endlessly for newer wants, as fast as they are satisfied." It would only take "advertising and other promotional devices" and "carefully predeveloped consumption" to build a "remarkable" momentum.[30]

An old-fashioned conservative like Leacock had little use for the apostles of progress and their pursuit of the "phantom of insatiable desires."[31] But for the liberals of the day—and ultimately the trade union movement and the evolving mass culture—the way to full employment lay along the aisles of new things called supermarkets and roads crowded with automotive traffic. Hoover's successor Franklin Roosevelt even moved Thanksgiving forward a week to make more time for Christmas shopping. Before World War II, radio was well established along with mass circulation magazines and newspapers. In the postwar period television was soon to place a fast-talking salesman in every living room. Buying became linked up with another looming pastime—watching.

Why? Why did consumerism "work," replacing labour's old demand to "work less and live more" with demands for money, not time? It might be tempting to attribute the trade of time for money to the nefarious influence of The Box. But that is unsatisfactory, confusing cause and effect. Freudian Marxist Herbert Marcuse echoed the Tory Leacock. Marcuse, an exile from Hitler's Germany, stayed on in the postwar United States and looked around at its explosive growth and attendant consumer culture. For Marcuse, capitalism had gained the ability to anaesthetize people, stimulating simple desires that could easily be fulfilled by market relationships. Soon, he argued in *Eros and Civilization*, the ability to produce more things with fewer people might make much work unnecessary, and free time could take over. "The result," he wrote hopefully, "would be a radical transformation of values. . . . Advanced industrial society is in permanent mobilization against this possibility."[32]

In 1930 breakfast cereal magnate W.K. Kellogg launched an experiment in decreased work-time, creating 25 per cent more jobs at his Michigan corn flakes plant by cutting the workday to six hours and adding a fourth shift. Like Stephen Leacock, he figured that a logical future would not bring infinite growth. Instead, technology would create more "free" time. The workers took no cut in pay. The experiment lasted over fifty years.

In 1932 a U.S. labour department Women's Bureau research team travelled to the Kellogg operation in Battle Creek and found that 85 per cent of the women workers liked the short shift. It meant they could pass more time with their families, relaxing and participating in self-activity like canning and organized games. Just after World War II management offered more money if the unionized workers would return to the eight-hour day, but the vote was three to one in favour of less work. From that time on a protracted struggle unfolded. On one side was a coalition of senior male workers and company managers who wanted more hours, and on the other a group of women workers committed to less work. Workers were allowed to choose workdays of six or eight hours.

After the 1950s, historian Benjamin Hunnicutt reports, the eight-hour advocates "abandoned the language of freedom and control that both men and women had been using for over 50 years, insisting that money was the only job benefit."[33] According to this perspective, work—not free time—was the centre of life. Those who wanted to work less were denounced by other workers as "silly" or "crazy" or "weak girls" or "lazy, sissy men." Those who continued to work the six-hour shift (three-quarters of them were women) called their opponents "work hogs" and defended "*our* shorter hours." They liked having more time outside the sphere of necessity, time free from both paid work and domestic chores, time to spend with kids, to go birding, to go to the park, to crochet—in short, time to do what they pleased.

Management eventually abandoned the gentler, human-relations technique of trying to persuade the Kellogg's women that work was all important. Instead the company picked up the stick. In the 1980s managers began talking about competitive pressures and threatened to relocate the jobs to places where the workers were more compliant. On December 11, 1984, a majority of the six-hour workers voted to end their "experiment."

The Kellogg experience provides a glimpse of how we have come to see leisure as an effortless time. The passive culture of consumption replaces other forms of activity. For Hunnicutt, the story of the corn flakes makers raises important questions: "Why go see the women play baseball when you can watch the Detroit Tigers on TV? Why do your own canning when you can buy canned goods at the supermarket? Why do anything in leisure time when you can pay someone else to do it? Why should I put in the trouble, why should I bother to expend all this effort when I've done my duty, I've put my time in at work, now I can just cease being a human being, put my brain on hold."[34]

Putting your brain on hold: the metaphor would have been foreign to Leacock or Marcuse—or to anyone who grew up in the early years of the telephone. After the war the idea of a reduction of labour time disappeared from the public agenda. Labour dropped its long-standing demands for shorter hours. And—for a brief historical interlude—the accommodation between labour and business brought low unemployment and wide-reaching prosperity and security—at least temporarily—to more people than had ever had it before.

Arguments for shorter work-weeks or fewer hours, however, do not get to the heart of the matter. While important, they are by definition qualitative. (As Lewis Mumford put it, "In time-keeping, in trading, in fighting, men counted numbers; and finally, as the habit grew, only numbers counted.")[35] To get at the heart of the matter, it is important to consider not just time but also *pastimes*. The various ways in which we pass our time reflect both our values and those of the culture that has spawned them, the same place where we learn those values. Do we pass our time passively? Do we squeeze time from busy work schedules for quickie power-vacations?

For my Catholic friend—the one who wondered about those stones lying out in a field on Sunday—capitalism's compulsive work/consume ethic is such a mess that it is impossible to simply "social-engineer our way out of it." Rather, the answer has to mirror the complexity of the dilemma, combining social and economic change with something he calls "spiritual discipline"—doing *without* to meet the needs of others, avoiding waste out of respect for nature or the poor, considering oneself a steward rather than the owner of things.

The values inherent in this rumination of time and spirituality are precisely those that Marcuse said industrial society was permanently mobilized against. They spring from a notion of love that is bound up with putting oneself at the service of others. *Service*: not in the sense of a service economy in which the word assumes the form of another commodity, where service means a new form of servitude, but in the form of stepping back and *thinking*—and not just in quantitative terms but in qualitative ways—searching, as Charles Taylor puts it, "for ways to recover a language of commitment to a greater whole."[36]

Mary Veley of Kingston has spent all of her adult life as a housewife and a good deal of it as a skein winder and heat setter in a yarn plant. Before that she worked in a knitting mill. Factory work finally led to back problems, and she had to quit work a few months before Kingston Spinners closed down to consolidate production in Quebec and Georgia.

"Hard work has never bothered me," she says. "I've done it all my life. Even with my back injury I didn't want to give up my job. But I was told if I stayed it would only get worse." Aside from factory work Mary Veley has had another working life. As the mother of three grown children she has been cook, cleaner, nurse, referee, psychologist, seamstress: Mother, Housewife. If love is all about putting oneself in the service of others, she has been in service as a caregiver all of her adult life.

Veley's mother arrived in Kingston from Quebec without being able to speak a word of English. Her father worked as a janitor at The Base. The area is lucky that CFB Kingston survived the military budget cuts of 1994. (Although 435 jobs disappeared at The Base, Bombardier's Kingston plant was revived by a contract to build transit vehicles for Malaysia.) Veley's son has his Grade 12 and a job driving a truck for a local steel fabricator. He says it is not something he wants to do all his life. Her nephew has a job at a packaging firm but has been applying everywhere else he can, "just in case . . . ." Her husband, a welder, has his own shop. Veley herself can rhyme off all the plants that are closed along Dalton Avenue in the industrial zone north of the CN main line.

Veley has thought a lot about what she could do now that the kids and the job are gone. Floral arranger and receptionist are a couple of jobs she thinks she could handle, jobs that would not be too hard on her back. She has been on Workers' Compensation for two years, with lots of time in physiotherapy, but her back still goes into regular and painful spasms. The Board wants her to get retrained so they can get her off their benefit rolls. So she has done the "Orientation to Employment" at St. Lawrence College. She has done the Job Finding Club at the March of Dimes on Patrick St. around the corner from her house. They did practice interviews and learned about resumés. The same week a newspaper story told about a local developer who had received three hundred applications for a receptionist job and was forced to put another ad in the classifieds telling people to stop sending in resumés.

But Veley has not yet fallen into the category that the labour-market policy types refer to as "discouraged." What's more, she's ready to play the training game. Her shining formica kitchen table has a neat stack of files with all the papers, pamphlets, and exhortations she's received since she began to participate in the great training tournament. The papers include her own notes, written in a neat hand on sheets of lined foolscap.

At the top of page one she has written "Places To Get Training For New Jobs" over a list of private business colleges. Her notes indicate two categories that she has been told are crucial to success in becoming a model employee in the 1990s. The first is "Human Relations." All the bosses these days say they want team players, so Veley has written down "getting along with others," "assertiveness," and "keeping attitudes." "At Kingston Spinners, I trained girls on the machines," Veley says. "You had to get along with others and take responsibility." But the second category is computer skills, and for her the world of dBase III plus is a different matter. She has noted the need to learn where all the keys are on a computer as well as "what some of the functions of the keys are." She is now on her third training course, this one at the Academy of Learning: computer familiarization, four hours each week for four weeks.

Like many other people who find themselves in this situation— and like many women in specific—Mary Veley has a fistful of skills that she herself doesn't recognize as such. These are things that, like so much else, she takes as given: tacit skills that are not necessarily available out there on the training market, abilities that you can't pick up off a shelf.

A few years back her daughter Laurie survived a hideous car accident and spent two months in a coma. After finally being released from the hospital, Laurie was partly paralysed. Veley took her home and began to work on her recovery. It was a long, slow, and painful process that involved teaching Laurie how to move her legs to walk and her hands to write. All the while Veley was working twelve-hour shifts at the yarn mill. After a three-month period of convalescence her daughter had recovered enough to get about on her own and look after herself.

(Advocates of a saner approach to working hours and labour standards in general argue that employees should have the right to take paid leave to deal with urgent family matters. The most common management response, at least in North America, is typified by Bruce McGillivray of Allegheny Ludlum, a steel company. "You can't take a day off because something's going on in your family," he told a *Wall Street Journal* reporter.)[37]

At one of the job-finding clubs, Veley was urged to add "experience as a caregiver" to her new resumé. But she only reluctantly gives details of this work. She feels it is just one of the things that fall under the job description "mother." She doesn't see it as any kind of formal credential. "I wasn't trained as a caregiver," she says frankly. "I looked after my own daughter, but I never looked after other people."

Now, when she reflects on this experience and her current situation, Veley is convinced she would like to find work caring for people. "Once you've been through something like that, you can really understand what people go through," she says. But with her back problem it would be difficult to work at hospital or home-care jobs that are physically demanding. She looks hard at the thick binder with all the papers from the courses she has taken, all the careful notes she's made. Her skills, her "really useful knowledge," have no apparent commercial value.

She is good at *caring*, in both literal and figurative senses of the word. In a world in which care is often sold like Jello or Nintendo, the skills she has can be salable. But when purchased—either through private home-care agencies or public facilities—they tend to be undervalued. Home-care workers hardly command rates of pay competitive with people who dream up advertising concepts for Nokia cell-phones. Indeed, with health-care cutbacks and increased reliance on out-of-hospital "community" care, those who do the looking after are more and more frequently women like Mary Veley. They do it on their own time, as part of their work as mothers, wives, daughters.

Nonetheless, Veley feels that she would like to work visiting seniors in their own homes, providing older women with hairdressing, light cleaning, and company. She once asked the WCB if they would help her out with hairdressing training, but her request was refused. So she is stuck trying to learn the rudiments of computer work. She finds it frustrating, but plays along.

"You need a certificate saying you have the skill," she says. "I did my mother's hair for years. I give my girlfriends perms and colouring. I began cutting my brother's hair after his barber retired, and he said he'd never go anywhere else. If it's something you're interested in, they should give you a chance at it."

The issue here is not simply just that Veley has been working all her life, wants to find work now, and finds herself with so many unused units of time. She is a caring person cast adrift in a world in which idle people and unfulfilled needs walk hand in hand. This is the mismatch that social policy should be attempting to address. It requires an alternative social vision, one that recognizes the existence of profound human needs that can't be served by the market. Unfortunately, people who find themselves without paid work will be quick-marched—as a condition of public provision—into whatever the private sector happens to have by way of low-wage work. Maybe they will be retrained, through some sort of learnfare program, for a job that may or may not exist. Or perhaps the jobless will be forced, like those convicted of minor crimes, to do "community service."

A more genuine and humane shift in social policy would involve a complete recasting of priorities, rooted in the recognition that the right to a decent income and the right to work should not necessarily be linked to a paid job in the conventional sense. This does not imply some pared-down version of the welfare state. A genuine and humane shift would speak to the need for a "policy of time."

"*Work*—or time exchanged for a wage—would no longer be one's principal occupation," French social theorist Andre Gorz says. "Everyone would—or could—define themselves with reference to their free time activities."

Such a future would defy the logic of competition and the war of each against all. The search for true "family values" would involve escaping capitalism's squirrel cage. It would mean a reduction in the activity of people-as-consumers, a break from the unsustainable notion of infinite growth on a finite planet—a break with the values of the market, values based on a belief that something's worth can ultimately only be determined by its price. The shift would recognize that the obscene disparities that fracture the planet between rich and poor represent the true meaning of globalization.

A "policy of time" would only be realistic if it were accompanied by a similar redistribution of material wealth (income), because we also live on a planet where too few have too much, and too many too little. According to Gorz:

> What would happen to the ethic of speed and punctuality, of 'we're not here for fun'—an ethic inculcated into children at school ever since the invention of manufactures? What would happen to the glorification of effort, speed and performance which is the basis of all industrial societies, capitalist and social-ist? And if the ethic of performance collapsed, what would become of the social and industrial hierarchy? On what values and imperatives would those in command base their authority?

Gorz points out that millions of bosses large and small recoil from such ideas. "Instinctively," he concludes, "they prefer unemployment to more free time. For unemployment is a disciplinary force."[38]

The glorification of effort and performance is not confined to the bosses. When trade union official Miriam Edelson found herself ground down by her time-gobbling, high-powered job and the need to look after an ailing child, she began to worry about her own health. She asked the Canadian Auto Workers union for time off: "I was told—point blank—that I wasn't committed enough." When the union—an organization explicitly devoted to the values of sharing and social justice—was less than sympathetic to her need for more time, she says she felt "betrayed."39

The real alternative to continuing unemployment is not just to break the fixed link between income and a *job*. It is to reconceive jobs so that they take on the more positive characteristics of work-as-vocation, the way Mary Veley sees caring for her injured daughter. We have seen a job as something that you "get" or "keep" or "lose." It has often been associated with criminals, whose slang describes a job as the next stick-up. A job has a limited sense, an occasional project, as in doing "small jobs" (or lumps) of work.40 It is often ordinary, too often negative—as in the sense of a Burger King or assembly-line or computer terminal. Jobs in a technologically advanced society, warned Senator David Croll in the wake of his inquiry into poverty, "make work as a means to any end other than putting food on the table and paying the bills, most uninviting."41

Work as *vocation* or *occupation* has a positive resonance, implying a combination of the mental and the manual, a reintegration of conception and execution, which the industrial division of labour has done so much to drive apart. Work can be inviting, something one does not for a *living* but to breathe fresh air into life—or to nurture life itself, the reflection of what could be called a *caring society*. We think of cultivating plants in a garden, an activity that differs starkly from the job of a field hand. It is only because a job implies a social relationship (usually hierarchical) that a woman who is nurturing children and managing a household—devoting much of her life to the servicing of those she loves—can be said to be *not working*.42

At age forty-six, after twenty-four years of stamping out bumpers, Dave Lachapelle found himself out of work, with a Grade 10 education and few apparent prospects. Things were changing all around him.

Lachapelle spent a year wondering what to do and realized that the age of falling expectations meant that he might be forced into work that paid half of what he used to make. He got odd jobs around Viscount Estates, where half the residents are senior citizens. Lachapelle, who describes himself as "sort of a half-assed handyman," put his Mr. Fixit skills to work doing the little bits of work that old people can't get commercial contractors to undertake—minor plumbing, panelling, painting, and roof coating.

After a year he decided to hop on the training bandwagon. High-school upgrading at St. Clair College allowed him to extend his UI claim, but he still had no idea about where he was headed. Then he remembered his first-ever job, working as a porter at Riverview Hospital, helping the patients get up in the morning, serving meals, feeding those who needed help and getting them to physiotherapy. He had liked that work, particularly in comparison to the clang, grit, and monotony of the auto-parts plant. He remembered how the patients appreciated the help others gave them.

His wife of twenty-five years was supportive during this period. Going back to high school can be a traumatic experience for people who have been independent and never felt that more education would ever be needed. This is particularly true of those like the Lachapelles, whose kids have already finished school and are now adrift in the tricky currents of the youth job market.

Dave Lachapelle completed another training course. This one gave him a health-care aide certificate and a chance at a job working at the Essex Nursing Home, a ten-minute walk from the Viscount Estates. Three days after he pocketed his certificate he found himself on call as a part-timer for the Reliacare Corporation. The hourly pay was less than he had got in his former full-time job, but at least it was something. Besides, he found that he was now doing really useful work. His transition from widget-maker to caregiver hints at a different sort of world of work, one that earmarks a Caring Society. The whole thing has something to do with human needs—both his own and those of others.

> When I worked at Windsor Bumper I hated the idea of just picking a bumper off the rack, putting it in the press, hitting the buttons, picking a bumper off the rack, putting it in the press, hitting the buttons. As far as the actual work goes, I'd have to say the nursing home is better. You're dealing with actual people and they appreciate what you're doing. If you're making bumpers for a company, they're making lots of money but they don't appreciate you. And I think a man needs that. You need the acceptance of other human beings. You need appreciation showed for what you're doing. It's part of being whole and remaining sane in a society which continually tries to drive you nuts.
>
> I was saying to my wife last night that there's this one lady, Stella. She never speaks. She sits in a chair and has this look on her face and says "Unhhh . . . unhhh . . . unhhh" all the time. I'm nice to her, like I try to be nice to everyone. I go up to her and she'll reach out and grab my hand and just hold it and look right in my eye and she'll just squeeze. For me things like that make work worthwhile. Makes you feel like something.

This Caring Society is one possibility. The Gambling Society is another—and one that governments across the country seem to be pushing as an underpinning of the new economy, a source of revenue and jobs. But it is uncertain as to whether the work spawned, for instance, by Windsor's new casino offers the kind of feeling Lachapelle talks about experiencing in his work as caregiver.

At its gala opening, trade-unionist-turned-cabinet-minister Frances Lankin described Canada's first Las Vegas-style casino as a "terrific jobs effort." Blackjack dealer, security guard, waiter: it was with no apparent irony that the head of Circus Circus Corporation (a casino partner) followed Lankin with the prediction that the newly created jobs would result in "a tremendous economic upheaval."[43]

The upheaval was the result of a remarkable social consensus in Windsor. Everyone from the Chamber of Commerce to the Labour Council jumped on the gambling bandwagon after Windsor had been eviscerated by a combination of recession, free trade, and a high dollar in the late 1980s and early 1990s. Gambling seemed to offer a return to prosperity. The province legalized casino gambling, and Canada's first big-time Las Vegas style casino opened in 1994 in the municipal art gallery, itself a converted brewery.[44] Sculptures and paintings were consigned to a suburban mall until a permanent building with more of the requisite glitz could be erected to stare across the river at Detroit. Most of the gamblers come from the United States.

Part of the debate (what there was of it—the development steamrollered a desperate city) over the introduction of the casino centred around the possible influence of organized crime and increased prostitution. Would the service jobs created by slot-machine tourism be able to replace the union jobs in the auto sector that had evaporated or emigrated? Such was the casino's momentum that this question didn't really seem to matter. Windsor apparently had no choice. There was simply no other way.

The city was ideally located for the same reason that it remains an important centre of car production, close to the American industrial heartland, not far from Chicago, Cleveland, and, of course, Detroit.

The United States has been called a "gambler's society."[45] One of the most attractive things about the country has always been the very real hope that everyone, no matter their origins, has a chance of hitting the jackpot, making it to the pinnacle of wealth and power. Even as that nation's relative economic power declines and it becomes stymied by its violence, poverty, and the anguished malaise of its modernity, people from all over the world still want to come and take a chance in the Gambler's Society. Each year Canadian newspapers feature ads for a "Green Card Lottery" that offers private services to help people with the annual U.S. immigration sweepstakes. They are attracted to the powerful American promise of freedom and opportunity, the chance to spin the wheel of fortune and come out a winner.

For Canadians—and particularly English Canadians in search of identity—this attraction to the United States remains ambivalent. We search for those things that separate us from Americans. At least on this side of the border an unlucky illness or accident—on the job or off—doesn't foreshadow a steady slide into poverty. For the Gambler's Society has its dark side. It is far easier to find yourself "out on the street"—both figuratively and literally—in the United States, a country whose employment regulations give much less weight to job stability than do those in most industrial societies. The poor in the United States have relatively little in the way of community support. A liberal politician like Bill Clinton favoured cutting people off welfare after two years. David Popenoe, the U.S. sociologist who coined the term "gambler's society," points out that to be well-off in his country means having both freedom and affluence; to be poor "is to be a second class citizen in a way that is not found to be acceptable" in many other lands.[46]

But Canada, too, is becoming more of a gambler's society. On the most obvious material level, we witness governments at all levels scrambling to maximize their tax takes, not by making the tax system fairer, but by bringing in new lotteries: 649s, Super-Lottos, Pro-line sports systems. Every mall offers its specialized lottery kiosk, the fantasies and realities of abundance. Gambling, whether it is a night at a smoky bingo hall in the Legion, a day at the races, or the big payoff from buying a ticket at the corner store, allows space for individualized fantasies. People with few choices or chances in the rest of their lives get the opportunity to make decisions that could make a real difference. (The Lotto 649 slogan is "Imagine the Freedom . . . .") Even if we never even participate, who among us hasn't caught ourselves imagining what we would do if we ever won The Big One?

We have more difficulty—even when robbed by unemployment or subemployment of the possibility of using our skills—of conjuring up the vision of an alternative based on the radical rethinking of how we spend our time and what we find rewarding.

So gambling, once confined to racetracks and shady back rooms, is now all around us. Churches that also protested Sunday shopping have raised moral objections, but few others seem to care. Gambling seems to respond to the apparently universal urge to take a chance, be a winner.

This is where the other side of the gambler's society shows itself. Canada, particularly English Canada, is becoming more and more like its southern neighbour. As in the United States, the abyss of poverty and despair that awaits the losers in the gambler's society yawns ever wider and deeper. Many more of us are poorer, more still insecure, as the market is left to sort the good job winners from the bad job (or no job) losers. Indeed, everyday jargon describes someone who is either poor and disreputable or just plain unsuccessful as a "loser." At the same time our public and political discourse is saturated with "win-win" solutions. There was a prime example of this in early 1995, when twenty-thousand people lined up at the Metro East Convention Centre in Toronto to apply for work on a General Motors assembly-line. One man waiting to fill out an application form reflected on the odds of getting a job: "Life's a chance, so why not take a chance on getting a job?"[47]

The process of Americanization is partly the result of a long historical trend that some Canadians have resisted but many, from the Liberal King to the Tory Mulroney, have accepted. It is the stuff of classical Canadian angst. In few other places do a people define themselves not by what they are but by what they are not: that is, American. By the 1990s Canadians grasping for the hallmarks of their own distinct society amidst the rising tides of continental integration often held up their social programs, particularly medicare, as an example. We have it. The Americans don't.

Put aside the shakiness of public health care in Canada, the U.S. administration's glacial (and apparently futile) moves in the opposite direction, and the largest province proclaiming a "win-win" situation when it succeeded in diverting a tiny portion of professional basketball profits to its hospitals. Put aside the fact that by the mid-1990s the main political opposition that English Canada had sent to Ottawa was touting an American-style individualism that simply wanted to get government out of the way and put the boots to the poor. The Red streak in Toryism had breathed its last with Brian Mulroney's callous cynicism.

Mutual affiliation and solidarity are foreign to the gambler's society, where it is more likely society-as-crapshoot. This tendency gets played out clearly in the labour market. Education and training are held out as the *sine qua non* for those who are to be winners. At the same time, both elements become more expensive and hence ever more the preserve of those fortunate enough to be what was once called "well born." So pervasive is the faith in learning and skill as the key to the future that the nagging question of who gets education and training and who does not is shunted aside. University tuition has doubled in the past ten years, with students paying a larger share of university costs. "These trends will no doubt continue," was the dry, apparently indisputable conclusion of the 1994 Liberal green paper on social policy.[48]

As a result, affluence has staged a comeback as a symbol of superiority. If you don't make it, if you do not survive, it is because you lack the skills that separate you from the winners. Those with the means treat schooling as another market, shopping around for the best schools in the best neighbourhoods—when they don't secede from the public system by joining the accelerating rush to private education. Then it is on to the university lottery as young people compete for admission to the ever more expensive schools at the top of *Maclean's* fervid annual ratings.

Windsor exemplifies the stresses of a working-class city with a rough egalitarianism based on trade union power in traditional manufacturing. More and more working-class children set their sights not on the University of Chrysler but on the University of Windsor as the key to the future. The fact that the city grasped the dubious promise of a gambling economy signals the stresses of a town that stares relentlessly across the Detroit River at the horrific wreckage of the American dream that is Detroit.

Kingston, for its part, gazes placidly across the charming blue waters of the St. Lawrence River and its Thousand Islands, which the city hopes will continue to attract touring yachts, affluent retirees, and the professional class of emerging information and bio-tech

industries. It stakes its future on an elite university that is in the vanguard of what has come to be called "privatizing" the programs it
offers, where $20,000 or more will buy you a business degree.
There's an "executive MBA" on offer. The trend is also known as "full-
fee" education. Big money will purchase access to programs that, we
are assured, promise *excellence.*

The wheel of fortune, of course, does much to determine one's
life chances. The kids from the right side of the tracks—in Kingston,
on the right side of Princess Street—will have a good chance at all
that learning. In Kingston the social divisions remain, reminding us
that we by no means live in a meritocracy. The wheel of fortune con-
tinues its relentless spin.

This is an undeniably dystopian vision, not perhaps as bleak as
George Orwell's look at the future in *1984,* but akin to it nevertheless. Remember, in *Coming up for Air,* Orwell's picture of modern life
as a frenzied struggle to sell things: "With most people it takes the
form of selling themselves . . . ."

In the final weeks of Orwell's life his old prep-school chum and
fellow writer, Cyril Connolly, wrote an article for the final issue of
Horizon, a literary magazine that had published some of Orwell's
finest work. Connolly must have been depressed by the end of the
magazine and the imminent death of his friend, whose last great
work was no cheery comment on modernity. "It is closing time in
the gardens of the West," he wrote, "and from now on an artist will
be judged only by the resonance of his solitude or the quality of his
despair."[49]

In the face of what seems to be overwhelming evidence, we must
search for a metaphorical counterweight to fortune's wheel. Time is
of the essence, and with a planet both divided and despoiled, this is
true in more ways than one. Connolly's words hint at an alternative
to the Gambler's Society. Good gardens take time, usually years, to
develop. This is especially true in a cold climate such as ours. Gardens are universally popular, appreciated by almost all cultures, the
inspiration for artists, the magic playground of children, the private
haunt of lovers, the symbol of sustenance.

Growing things for the love of growing things is hard to rival as a pastime. While so many other pursuits are governed by a burgeoning expertocracy, gardening is ideally—and most frequently—the preserve of the amateur. It is not just for large landowners. Gardens are democratic, because anyone interested in cultivation can care for a small plot—perhaps an allotment with tomatoes and greens, an apartment balcony teeming with growth, a simple backyard plot. Caring for plants and flowers is physical work, and digging and weeding are good for the body as well as the soul.

Gardening is the kind of work that stands in glaring contrast to so many tasks regulated by the clock and the other machines that so dominate working lives and leisure-time activities. Indeed, mechanization and the market have trouble subverting this pastime, although oddly enough some people seem to prefer the shrill whine of the leaf-blower to the quiet brushing of the rake.

The work of gardening is usually not a job. It is more often a hobby, a diversion from life's other activities, a form of caring and nurturing. Its built-in soulful character distinguishes its labour from much of the rest of the way we pass our time, in work or in leisure. If someone says they love television it might be a little hard to take them seriously, or at least difficult to empathize with their passion. But most of us nod with understanding when people talk of their great affection for their roses. This is all bound up with caring, as opposed to gambling.

The distinction of the organic from the mechanical is part of a long tradition that goes back to Coleridge and Thoreau. It is possible—and sometimes even dangerous—to overplay organic metaphors by applying them to whole societies, rationalizing hierarchy and some supposed natural order of things. But the idea still has resonance, as it did for poet James Oppenheim in 1912 when he noticed a banner carried by a Massachusetts woman. She was one of twenty thousand workers who had walked away spontaneously from their mill jobs to protest an arbitrary cut in pay. Her sign bore the simple message: "We want bread and roses too."

From the sight of the women "marching in the beauty of the day" sprouted a line of verse: "Hearts starve as well as bodies; give us bread but give us roses."

Training is in good measure about ratcheting down the hopes and expectations of *people*, who become "human resources" to be called forth and filled up again and again with the skills now demanded, now declared obsolete, by a fast-changing labour market—and this is where "hearts" come in. Demands for less work-time must be rooted not just in the need for jobs, jobs, jobs. They must also have a cultural dimension, for without that dimension the issue of less work (a "policy of time") is unlikely to be resonant enough to spark the popular imagination.

It is true that the urge to take a chance, to compete, to be a winner is a powerful one. If we are to move from a gambler's society to a caring society, we must recognize Leacock's "phantom of insatiable desires" for what it is: a ghost of a chance based on capitalism's need to expand—forever. Only a relative few, both within Canada and around a world torn by unshared bread, will emerge as "winners." The odds are stacked, the game fixed. If we fail to challenge the mania of the gambler's society, most of us—and the earth we live on—can only be losers culturally or materially, or both.

All of the above thoughts spring out of what I learned from working-class people, from what I found in the archives or read in books, and (need I say it?) from my own prejudices and experiences. I have never really been a gambler. I do not understand the intricacies of scratch 'n' win at the corner variety store. Like many other dutiful Canadians, I save carefully and do not take risky chances with my RRSPs. I admit to being somewhat unnerved when I get a mailing from an investment dealer telling me to "forget about" public pensions: "There can be no exceptions to this rule." Despite the warning, I will continue to play it safe.

Nor am I much of a gardener. Although I will put in a few practical kitchen herbs come May each year, I have never developed the passion for the garden exhibited by many people I know, including my mother Olive and her grandmother before her. (Is the garden more of a female pursuit? Certainly gardens on farms were traditionally tended by women.) Even so, I have enjoyed working with my mother in her handsome rock garden in recent years, when her health has prevented her from doing the heavy work. I would happily dig new beds according to her specifications.

Then, in the middle of the time when I was writing this book, my mother became seriously ill, and was confined to various hospitals, in need of various levels of care (intensive, acute, long-term . . .). It became apparent to me that she would never again be able to look after her precious garden. A woman who had devoted her own life to volunteering her time in churches and libraries and, in general, caring—for her children, her own parents and her mother-in-law, her husband—was now herself in need of care.

In one hospital after another—four of them—my mother received the best of care from nursing and housekeeping staffs whose ranks have been thinned by public-spending cuts. When she left one place she would keep in touch with her former caregivers, scrawling out notes in her shaky hand. I am sure that one of the reasons she fared so well in these institutional settings was that she was so thankful for receiving any bit of routine care that the gratitude became reciprocal, making work that is becoming more pressured and subject to speed-up just a bit more enjoyable, seemingly more worthwhile.

The situation reminds me of something that Dave Lachapelle told me when he was describing what it was like to be working in a home for the aged instead of in an auto-parts plant. "Older people get neglected, so they appreciate anything you can do for them. I sorta get a good sense out of it. If you go in just saying, 'This job pays fifteen bucks an hour, it's great . . .' you're fooling yourself. All that's gonna happen is you'll turn bad."

Unlike gambling, an individualistic and often compulsive pursuit, caring—and a caring society—is about choosing to be connected, to be involved. It reflects something mutual, something that Orwell would have described with that favourite word of his, *decency*. In 1946 Orwell wrote that the important issue of the day was not whether "the people who will wipe their boots on us during the next fifty years" would be labelled managers or politicians or bureaucrats. The question was whether capitalism, doomed in Orwell's eyes, would give way to oligarchy or true democracy.

Capitalism is still with us today, with a vengeance, so Orwell was certainly wrong on that score. But in the same essay, a discussion of James Burnham's book *The Managerial Revolution* (a major influence on *1984*), Orwell anguished over the tendency to what he called "realism"—the tendency to assume that the thing that is happening now will simply continue. He called this inclination not just a bad habit but a "major mental disease."[50]

Those of us who, fifty years later, would imagine a Caring Society of connectedness, of decency, are accustomed to being told to "get real" or to reconcile ourselves to "the real world." But we must recognize this talk for what it is. It is not just sturdy pragmatism that can be juxtaposed to some dreamy cloudland inhabited by idealists. It is, rather, what Orwell would have called a "smelly little orthodoxy."

Today's "real world" is the world of the market, a place in which equality and our cherished individual freedoms are overwhelmed. The economic lives of our cities and the people in them are now dominated by a vast global market that apparently cannot be controlled nationally, let alone locally. This is not a world in which the needs of people who are really in need—in Windsor or Kingston, in Bangladesh or Guatemala—have any priority. Orthodox wisdom has it that this state of affairs is cast in stone—that we can do nothing but adapt to this new normalcy. The vision of a Caring Society recognizes that the problem with "normal," to paraphrase Bruce Cockburn's refrain, is that it "always gets worse." We can, surely, do better.

NOTES

CHAPTER I
Introduction: The Age of Falling Expectations

1. *The Globe and Mail,* January 9, 1995.

2. *The Gazette* (Montreal), December 24, 1994.

3. Economic Council of Canada, *Good Jobs, Bad Jobs: Employment in the Service Economy,* Ottawa, 1990, p.i.

4. Government of Canada, *The Canadian Pulp and Paper Industry: A Focus on Human Resources,* Ottawa, 1993, p.71.

5. Economic Council of Canada, *Employment in the Service Economy,* Ottawa, 1991, p.i, tables 2-1, 4-2.

6. T. Ran Ide and Arthur J. Cordell, "The New Tools: Implications for the Future of Work," in Armine Yalnizyan, T. Ran Ide, and Arthur J. Cordell, *Shifting Time: Social Policy and the Future of Work* (Toronto: Between the Lines, 1994), pp.106-7.

7. *The Toronto Star,* April 9, 1983; *The Globe and Mail,* December 5, 1992.

8. *The Wall Street Journal,* March 4, 1992, cited in Ide and Cordell, "The New Tools," p.109; Communications, Energy and Paperworkers (Craft and Services and Operator and Dining Services), submission to Bell Canada, September 30, 1993; *The Globe and Mail,* January 14, 1994.

9. Andre Gorz, *Farewell to the Working Class: An Essay on Post-Industrial Socialism* (Boston: South End Press, 1982), p.67. See also Gorz, *Critique of Economic Reason* (London: Verso, 1989); and Ivan Illich, *Tools for Conviviality* (New York: Harper Colophon, 1973), and *Shadow Work* (Boston: M. Boyars, 1981).

10. Fernand Braudel, *Civilization and Capitalism, 15th-18th Century*, vol.II, *The Wheels of Commerce* (London: Fontana, 1985), p.52.

11. Economic Council of Canada, *Good Jobs, Bad Jobs*, p.1; emphasis added.

12. J. Edward Newall, quoted in *The Whig-Standard* (Kingston), February 10, 1992.

13. Robert Reich, *The Work of Nations: Preparing Ourselves for 21st Century Capitalism* (New York: Knopf, 1991), p.109.

14. *The Globe and Mail*, May 4, 1993.

15. Economic Council of Canada, *Good Jobs, Bad Jobs*, pp.14, 15, 17. *The Gazette*, April 3, 1993.

16. Economic Council of Canada, *The New Face of Poverty: Income Security Needs of Canadian Families*, Ottawa, 1992, pp.5, 54.

17. John Myles, "The Expanding Middle: Some Canadian Evidence on the De-skilling Debate," *Canadian Review of Sociology and Anthropology*, 25,3 (1988).

18. John Kenneth Galbraith, *The Culture of Contentment* (Boston: Houghton Mifflin, 1992), pp.15-16, 27, 31, 38.

19. Author's communication with Robert Fisher, Corrections Canada, Kingston, November 1, 1993.

20. Economic Council of Canada, *Employment in the Service Economy*, p.137.

21. John Myles, Garnett Picot, and Ted Wannell, "Does Postindustrialism Matter? Evidence from the Canadian Experience," in *Changing Classes*, ed. Gosta Esping-Andersen (London: Sage, 1993).

22. Economic Council of Canada, *Good Jobs, Bad Jobs*, p.12.

23. Canadian Chamber of Commerce, *Putting Business into Training: A Guide to Investing in People: Focus 2000*, report of the Task Force on Education and Training, Ottawa, n.d., pp.2,8.

24. Kathryn McMullen, Norm Leckie, and Christina Caron, *Innovation at Work: The Working With Technology Survey, 1980-91* (Kingston: Industrial Relations Centre [IRC] Press, Queen's University, 1993). The Working with Technology Survey, conducted by the Economic Council of Canada, was done in two phases: 1980-85, and 1986-91, with a nationwide sample of several hundred establishments. The results were published at Queen's University after the Economic Council was closed down.

25. Economic Council of Canada, *Good Jobs, Bad Jobs*, p.24.

26. Gordon Betcherman, Kathryn McMullen, Norm Lecki, and Christina Caron, *The Canadian Workplace in Transition: The Final Report of the Human Resource Management Project* (Kingston: IRC Press, Queen's University, 1994).

27. D.W. Livingstone, "Enterprise Restructuring and New Training Programs: The Challenge for Labour," presentation to Conference on Training, Ontario Federation of Labour, October 1990.

28. David Noble, *Progress without People: New Technology, Unemployment, and the Message of Resistance* (Toronto: Between the Lines, 1995), p.110.

29. For a scholarly analysis, see Stephen McBride and John Shields, *Dismantling a Nation: Canada and the New World Order* (Halifax: Fernwood Publishing, 1993); for a more popular treatment, see Linda McQuaig, *The Wealthy Banker's Wife: The Assault on Equality in Canada* (Toronto: Penguin, 1993); for an analysis of women, poverty, and the job market, see M. Gunderson and L. Muszynski, *Women and Labour Market Poverty* (Ottawa: Canadian Advisory Council on the Status of Women, 1990).

30. C. Wright Mills, *White Collar: The American Middle Classes* (New York: Oxford University Press, 1951), p.xv.

CHAPTER 2

Windsor: "You Could Quit at Ford's in the Morning…"

1. City of Windsor, *Annual Report of the Assessment Commissioner*, Windsor City Archives, RG3, E1-1/9, various years.

2. *The Windsor Star*, July 25, 1967.

3. Quoted in W. Roberts and J. Bullen, "A Heritage of Hope and Struggle: Workers, Unions and Politics in Canada, 1930-1982," in *Modern Canada 1930-1980s: Readings in Canadian Social History*, vol.5, ed. M.S. Cross and G. Kealey (Toronto: McClelland and Stewart, 1984), p.116.

4. *The Globe and Mail*, January 31, 1946, quoted in D. Moulton, "Ford Windsor 1945," in *On Strike: Six Key Labour Struggles in Canada, 1919-1949*, ed. I. Abella (Toronto: James, Lewis & Samuel, 1974), p.149.

5. M.J. Piore and Charles F. Sabel, *The Second Industrial Divide: Possibilities for Prosperity* (New York: Basic Books, 1984), p.80.

6. Benjamin Kline Hunnicutt, "The End of Shorter Hours," *Labour History*, 25,3 (1984).

7. *The Windsor Star*, October 8, 1992.

8. B. Hamper, *Rivethead: Tales from the Assembly Line* (New York: Warner Books, 1986), p.40.

9. *Tri-City Labour Review*, April 13, 1932, quoted in Stuart Ewen, *Captains of*

Consciousness: Advertising and the Social Roots of the Consumer Culture (New York: McGraw-Hill, 1976), p.11.

10. Antonio Gramsci, *Selections from the Prison Notebooks*, ed. Quintin Hoare and Geoffrey Nowell Smith (New York: International Publishers, 1971). Gramsci identified the organization of human labour as no less a machine than the actual assembly-line and saw that high wages offered the possibility of a standard of living "adequate to the new methods of production." He was at pains to point out that although "the whole Fordian ideology of high wages" emerged in part from the needs of modern industry, it was not a "primary phenomenon" (p.311).

11. Windsor/Chatham Canada Employment Centres, Labour Market Information Department, "Labour Market Information," mimeo, March 25, 1993.

12. *The Globe and Mail,* March 17, 1993.

13. Sue Craig, "Playing Hardball With Tomatoes," *Financial Post Magazine,* November 1994.

14. Armine Yalnizyan, "Defining Social Security, Defining Ourselves: Why We Need to Change Our Thinking Before It's Too Late," Canadian Centre for Policy Alternatives/Social Planning Council of Metropolitan Toronto, 1993, p.6. See also Yalnizyan, Ide, and Cordell, *Shifting Time.*

15. *The Windsor Star,* April 16, 18, 1992.

16. *The Globe and Mail,* June 24, 1993.

17. R.K. Kunau and O.L. Crocker, "WESTAC/CEC Labour Market Survey," Windsor, 1993.

18. Yalnizyan, "Defining Social Security," p.4.

19. *The Toronto Star,* February 15, 1992.

CHAPTER 3

Kingston: "…With Not Much in Between"

1. N.F. Morrison, *Garden Gateway to Canada: 100 Years of Windsor and Essex County 1854-1954* (Windsor: Herald Press, 1954), p.vii.

2. Charles Dickens, "American Notes" [London, 1846], in *Kingston! Oh Kingston! An Anthology,* ed. A.B. Smith (Kingston: Brown and Martin, 1987), p.348.

3. John George Lambton, 1st Earl of Durham, "The Report and Dispatches" [London, 1839], in *Kingston! Oh Kingston!,* ed. Smith, p.301.

4. A.R.M. Lower, "The Character of Kingston," in *To Preserve and Defend: Essays on Kingston in the Nineteenth Century*, ed. G. Tulchinsky (Montreal and Kingston: McGill-Queen's University Press, 1974), pp.19-20.

5. R. Harris, *Democracy in Kingston: A Social Movement in Urban Politics, 1965-70* (Toronto: University of Toronto Press, 1988), p.79.

6. Lower, "Character of Kingston," p.30.

7. Harris, *Democracy in Kingston*, p.35.

8. Kingston Area Economic Development Commission, *Economic Development Strategy, 1990-2000* (Kingston, 1990).

9. Stephen Chait Consultants Ltd., "City of Kingston Industrial Screen Analysis Final Report," prepared for the Corporation of the City of Kingston, June 1992.

10. M. Yeates, "The Enhancement of Kingston," *The Whig-Standard*, February 15, 1992.

11. Author's communication with Dick Bowman, Queen's information officer, November 11, 1993.

12. Kingston Area Development Economic Commission, *Annual Report*, Kingston, 1993; *The Whig-Standard*, July 20, 1993.

13. QL Systems, "QL Systems Limited—A Summary," mimeo, Kingston, 1993.

14. J. Weizenbaum, *Computer Power and Human Reason: From Judgment to Calculation* (San Francisco: Freeman, 1976), p.31.

15. *The New York Times*, November 14, 1993.

16. Statistics Canada, *Survey of Literacy Skills Used in Daily Activities*, Ottawa, 1990.

17. George Grant, *Lament for a Nation: The Defeat of Canadian Nationalism* (Ottawa: Carleton University Press, 1982), p.49.

18. J.L. Granatstein, *The Ottawa Men: The Civil Service Mandarins 1935-1957* (Toronto: Oxford University Press, 1982), p.32.

19. Reich, *Work of Nations*, pp.177-78.

20. A. Ormiston et al., "Youth Unemployment and Underemployment in North Kingston: Report on a Study," mimeo, School of Urban and Regional Planning, Queen's University, 1988.

21. Insight Canada Research, "Aspirations Project Qualitative Research Report," prepared for the Premier's Council on Health, Well-being and Social Justice, Toronto, February 1993.

22. *The Globe and Mail*, August 28, 1993.

CHAPTER 4
The Training Gospel and the Army of Servants

1. N.F. Dupuis, "The Conservative and the Liberal in Education," *Queen's Quarterly*, October 1901, January 1902.

2. John Watson, "The University and the Schools," in *Queen's Quarterly*, April 1901, p.337.

3. Dupuis, "The Conservative and the Liberal in Education."

4. Parliament of Canada, *Report of the Royal Commission on Industrial Training and Technical Education*, testimony of George Howell (Toronto, October 1910), Sessional Paper No.191d, Ottawa, 1913, p.2088.

5. Quoted in Canada, *Report of the Royal Commission*, p.2087.

6. E.P. Thompson, *The Making of the English Working Class* (Harmondsworth, England: Penguin, 1968), ch.11.

7. Adam Smith, *An Inquiry into the Nature and Causes of the Wealth of Nations* [1776] (New York: Collier, 1937), p.10.

8. William Lyon Mackenzie King, diary entry, Nov.10-26, 1909, quoted in R.M. Dawson, *William Lyon Mackenzie King: A Political Biography*, vol.1, *1874-1923* (Toronto: University of Toronto Press, 1958), p.201.

9. Canada, *Report of the Royal Commission*, pp.9, 15, 19.

10. Ibid., p.24.

11. Ibid., p.2156.

12. Ibid., p.173.

13. D.S. Landes, *The Unbound Prometheus: Technological Change and Industrial Development in Western Europe from 1750 to the Present* (London: Cambridge University Press, 1969), p.41.

14. Ibid., p.63.

15. *The Globe and Mail*, November 25, 1993. The same news article also mentioned that the Canadian Imperial Bank of Commerce had announced a cut of twenty-five hundred jobs.

16. *The Globe and Mail*, December 27, 1945.

17. J.M. Pigott, "Are Employers Lethargic about Apprenticeship?" speech, August 30, 1953, Department of Labour Papers, Industrial Training Branch, Correspondence of Director, Archives of Ontario, RG 7 series 79 box 28.

18. Quoted in Harry Braverman, *Labor and Monopoly Capital: The Degradation of Work in the Twentieth Century* (New York: Monthly Review Press, 1974), p.131.

19. Canada, *Report of the Royal Commission*, p.19.

20. J.H. Thompson and A. Seager, *Canada 1922-1939: Decades of Discord* (Toronto: McClelland and Stewart, 1985), p.156.

21. Canada, *Report of the Royal Commission*, p.2341.

22. Ibid., p.2160.

23. Thompson and Seager, *Canada*, p.154.

24. Craig Heron, *The Canadian Labour Movement: A Short History* (Toronto: Lorimer, 1989), p.59.

25. J. Stefan Dupré, David M. Cameron, Grahame H. McKechnie, and Theodore B. Rotenberg, *Federalism and Policy Development: The Case of Adult Occupational Training in Ontario* (Toronto: University of Toronto Press, 1973).

26. Quoted in R.M. Stamp, *The Schools of Ontario, 1876-1976* (Toronto: University of Toronto Press, 1982), p.84.

27. Letter from Fred Hawes to A.W. Crawford, Acting Director of Training, Department of Labour (n.d.), December 1951, Archives of Ontario, RG7.79.26.

28. Pigott, "Are Employers Lethargic about Apprenticeship?"

29. A.W. Crawford to F.J. Hawes, July 10, 1952, Archives of Ontario, RG7.79.26.

30. Canada, Department of Labour, "Record of Approval under the Vocational Training Co-ordination Act," Ottawa, June 27, 1956, Archives of Ontario, RG7.79.26.

31. Department of Labour Vocational Training Branch, "A Report on Technical and Vocational Training in Canada," March 16, 1961, Archives of Ontario, RG7.12.77.4152.

32. Stamp, *Schools of Ontario*, pp.194-95.

33. Department of Labour Vocational Training Branch, "Report on Technical and Vocational Training."

34. Quoted in Alan G. Green, *Immigration and the Postwar Canadian Economy* (Toronto: Macmillan, 1976), p.21.

35. Ibid., Table 5-8, p.160.

36. Reg Whitaker, "Canadian Immigration Policy since Confederation," pamphlet, Canadian Historical Association, Ottawa, 1991, p.15.

37 "Minutes of the Apprenticeship Training Advisory Committee," Ottawa, October 14-15, 1954, Archives of Ontario, RG7.79.6.

38. F.J. Hawes to A.W. Crawford, July 14, 1952, Archives of Ontario, RG7.79.26.

39. G.H. Simmons to A.W. Crawford, December 12, 1955, Archives of Ontario RG.7.79.6.

40. D.C. McNeill to Hon. W.K. Warrender, January 29, 1962, Archives of Ontario, RG7.12.36.

41. F.J. Hawes to A.W. Crawford, January 8, 1953, Archives of Ontario, RG7.79.26.

42. Wolfgang Streeck, "Training and the New Industrial Relations: A Strategic Role for Unions," in *Economic Restructuring and the Emerging Pattern of Industrial Relations*, ed. S.R. Sleigh (Kalamazoo, Mich.: W.P. Upjohn Institute, 1993), p.179; emphasis added.

43. J.F. Dwyer to F. Hawes, January 8, 1952, Archives of Ontario, RG7.79.26.

44. CMA circular, January 7, 1966, Archives of Ontario, RG7.12.51.3245/2.

45. The 1961 census data are cited in G.R. Horne et al., *A Survey of Labour Market Conditions, Windsor, Ontario, 1964: A Case Study* (Ottawa: Economic Council of Canada, 1965), pp.14, 26.

46. J. Tingle to H.E. Lucas, January 28, 1969, Archives of Ontario, RG7.79.38.

47. Canada, *Report of the Royal Commission*, p.5.

CHAPTER 5

Public Purpose, Know-how, and the War of Production

1. Correspondence from M. Lett and case report, Deputy Minister's files, Archives of Ontario, RG7.12.51.3239.

2. Canada, *Report of the Royal Commission*, pp.174-76, 365.

3. Ibid., p.375.

4. Alice Kessler-Harris and Karen Sacks, "The Demise of Domesticity in America," in *Women, Households and the Economy*, ed. B. Lourdes Beneria and Catharine Stimpson (New Brunswick, N.J.: Rutgers University Press, 1987); see also Janice Acton, ed., *Women at Work: Ontario 1850-1930* (Toronto: Canadian Women's Educational Press, 1974).

5. Mills, *White Collar*, p.63.

6. Department of Labour Vocational Training Branch, "Report on Technical and Vocational Training," Archives of Ontario, RG7.12.77.4152.

7. Director of Trade Schools, F.W. Ward to Elma Avery, February 20, 1952, Archives of Ontario, RG2.P-3.420-3.

8. D.C. McNeill to J.B. Metzler, December 27, 1961, Archives of Ontario, RG7.12.36DC.

9. For details on TVTA, see Dupré et al., *Federalism and Policy Development*.

10. Ibid., p.18.

11. Tom Kent, *A Public Purpose: An Experience of Liberal Opposition and Canadian Government* (Kingston and Montreal: McGill-Queen's University Press, 1988), p.400.

12. "The Industrial Training Program of the Ontario Department of Labour" (n.d.), 1966, Archives of Ontario, RG7.12.51.3245.3.

13. Ibid.

14. Kent, *A Public Purpose*, p.361.

15. Kurt Vonnegut, *Player Piano* (New York, 1952), p.63.

16. Kent, *A Public Purpose*, p.357; emphasis added.

17. Leon Muszynski, "The Politics of Labour Market Policy," in *The Politics of Economic Policy*, ed. Bruce Doern, Collected Research Studies, Royal Commission on the Economic Union and Development Prospects for Canada, vol. 40 (Toronto: University of Toronto Press, 1986), p.262. Interestingly, the Macdonald Commission—which provided the ideological rationale for the Free Trade Agreement—was commissioned by the Trudeau Liberals, who felt the need for long-term thinking on the economy. The Economic Council of Canada and its intended function were bypassed and, even though it backed free trade, the Council was eventually abolished by the Mulroney Tories, who were simply moving harder and faster in the direction that mainstream liberalism had been taking for nearly a century.

18. Quoted in Muszynski, "Politics of Labour Market Policy," p.205.

19. Select Committee on Apprenticeship and Training, Terms of Reference, May 16, 1962, Archives of Ontario, RG7.12.77.4152.

20. Deputy Minister of Labour, files on automation conference, Archives of Ontario, RG7.12.38.

21. Ontario Department of Labour, "The Stationery Engineer in the Province of Ontario: A Socio-economic Profile, with Reference to the Training Requirements for Stationary Engineers," November 26, 1968, Archives of Ontario, RG7.64.8.

22. "Organized Training in Four Industries in Ontario," March 27, 1968, Archives of Ontario, RG7.79.38.

23. Tom Kent, speech to Institute of Public Administration of Canada, September 7, 1967, Archives of Ontario, RG7.12.55.3422.

24. Dupré et al., *Federalism and Policy Development*, p.60.

25. Muszynski, "Politics of Labour Market Policy," p.267.

26. Special Senate Committee on Poverty, *Poverty in Canada*, Ottawa, 1971, p.vii.

27. Ibid., p.xxxi.

28. "A Comparison of Selected Characteristics of Persons Authorized to Take Full-Time Training under the General Purchase Agreements in 1968-69 and 1969-70," Department of Manpower and Immigration, Planning and Evaluation Branch, February, 1970, Archives of Ontario, RG7.130.13.

29. Dupré et al., *Federalism and Policy Development*, p.117.

30. "A Comparison of Selected Characteristics of Persons."

31. Dupré et al., *Federalism and Policy Development*, p.126.

32. Economic Council of Canada, *Annual Review*, 1971, cited in Dupré, *Federalism and Policy Development*, p.125.

33. Dupré, *Federalism and Policy Development*, p.125.

34. "A Comparison of Selected Characteristics of Persons"; emphasis added.

35. Dupré, *Federalism and Policy Development*, p.123.

36. Westcott papers, Archives of Ontario, F2094-3-2-051. The training of child-care workers was initiated in Ontario in 1966 not by industry or government but by the Ontario Welfare Council.

37. Kent, *A Public Purpose*, pp.431-32.

38. Peter Warrian, "Macro Choices, Micro Failures: Perspectives on Economic Restructuring," presentation to Canadian Labour Market and Productivity Centre, Ottawa, May 1991. Warrian was a key actor in the establishment of a labour-management sectoral training council in the steel industry.

39. Advisory Council on Adjustment, *Adjusting to Win*, Ottawa, 1989.

40. National Advisory Board on Science and Technology, *Human Resource Development Committee Report*, Ottawa, 1991.

41. *The Toronto Star*, October 26, 1990.

42. See Kari Dehli, "Subject to the New Global Economy: Power and Positioning in Ontario Labour Market Policy Formation," *Studies in Political Economy*, 41 (Summer 1993).

CHAPTER 6

Private Troubles in the Border City

1. Lloyd Axworthy, speech to Women and the Labour Market Conference, OECD, April 17, 1980, Archives of Ontario, RG32.30 (Interim Box 008).

2. For a comprehensive analysis of training and changing economic structures, see Mechthild U. Hart, *Working and Educating for Life: Feminist and International Perspectives on Adult Education* (London: Routledge, 1992).

3. D. Stoffman, "Working Class Heroes," *Report on Business Magazine*, December 1993.

4. *The Windsor Star*, December 21, 1978.

5. *The Globe and Mail*, December 27, 1978.

6. "Publicly Reported Skill Shortages," 1978, Archives of Ontario, RG32.30.0008.

7. *The Windsor Star*, December 21, 1978.

8. For a comprehensive discussion of the shortcomings of the training solution to unemployment and economic development, see Derek Aldcroft, *Education, Training and Economic Performance* (Manchester: Manchester University Press, 1992).

9. Whitaker, "Canadian Immigration Policy Since Confederation," pp.19-20.

10. Muszynski, "Politics of Labour Market Policy," p.273.

11. B. Cullen to Thomas Wells (n.d.), received September 5, 1978, Archives of Ontario, RG32.30.008.

12. Canada Employment and Immigration Commission, "Labour Market Policies," discussion paper presented at First Ministers' Conference on the Economy, November 27-29, 1978, quoted in Muszynski, "Politics of Labour Market Policy," p.44.

13. Memo from T.P. Adams to W.F. Davy, E.L. Kerridge, H. Noble, June 29, 1979, Archives of Ontario, RG32.30.12.

14. Statistics Canada, *Labour Force Survey*, December 1986, Cat. 71-001.

15. House of Commons, *Work for Tomorrow: Employment Opportunities for the Eighties*, Ottawa, 1981, p.61.

16. See Muszynski, "Politics of Labour Market Policy," pp.275, 283.

17. Jennifer Stephen, "Training and Unemployed Workers," paper presented to What's Training Got to Do with It? conference, York University Centre for Research on Work and Society, June 10, 1993.

18. Rianne Mahon, "Adjusting to Win: The New Tory Training Initiative," in *How Ottawa Spends: Tracking the Second Agenda*, ed. K.A. Graham (Ottawa: Carleton University Press, 1990), p.79.

19. Health and Welfare Canada, "Canadian Paper on New Orientations for Social Policy," presentation to the Ministerial Meeting of the Employment, Labour, and Social Affairs Committee on Social Policy, OECD, Paris, December 8-9, 1992.

20. *The Globe and Mail*, January 16, 1993.

21. C. Wright Mills, *The Sociological Imagination* (New York: Evergreen, 1961), pp.8-9.

22. Health and Welfare Canada, "Canadian Paper on New Health Orientations"; Human Resources Development Canada, "Agenda: Jobs and Growth: Improving Social Security in Canada," discussion paper summary, Ottawa, 1994, p.23.

23. Robert M. Campbell, *The Full-Employment Objective in Canada, 1945-85: Historical, Conceptual and Comparative Perspectives* (Ottawa: Economic Council of Canada, 1991), p.14.

24. CEIC, *How to Find a Job* (Ottawa: Ministry of Supply and Services, 1991).

25. Ronald Dore, *The Diploma Disease: Education, Qualification and Development* (Berkeley and Los Angeles: University of California Press, 1976), pp.8-9.

26. Quoted in Dore, *Diploma Disease*, p.9.

27. *The New York Times*, March 25, 1994.

28. Stephen, "Training and Unemployed Workers."

29. Nicholas Saul, "'Organizing the Organized': The Canadian Auto Workers' Paid Educational Leave Programme," M.A. dissertation, Department of Sociology, University of Warwick, 1994.

CHAPTER 7
Collars of Many Colours in the Limestone City

1. Archives of Ontario, RG32-30, Interim Access Box 016. Archives of Ontario, RG32-30, Interim Access Box 0006, indicates that in 1979 there were twenty-three CITCs in Ontario; the Toronto CITC had eight industry and one labour representative; Kitchener-Waterloo had ten people from industry and none from labour; Hamilton eight from industry and one from labour; Sarnia/London eight from industry and none from labour. All had involved educators. The same files show that when skill shortages loomed, history could still repeat itself: "Area: Brockville College: St Lawrence Shortage of machinists identified. Comments: Industry solved their immediate problem by bringing in offshore, high-skilled labour on Special Federal Govt Permits—long range problem not solved."

2. Author's communication with Sue Bolton, jobsOntario Training Fund, January 25, 1994.

3. Information from jobsOntario Training Fund, County of Frontenac, December 2, 1993.

4. *Partners: News and Information for Our Partners in Learning*, 1,2 (1993).

5. *The Whig-Standard*, August 29, 1993.

6. Bertrand Russell, *In Praise of Idleness and Other Essays* (London: Allen & Unwin, 1963), p.11.

7. McMullen, Leckie, and Caron, *Innovation at Work*.

8. *The Whig-Standard*, January 13, February 9, 1994.

9. Canada-British Columbia Labour Force Development Agreement, pp.29-30, quoted in J. Calvert and L. Kuehn, *Pandora's Box: Corporate Power, Free Trade and Canadian Education* (Toronto: Our Schools/Ourselves, 1993), pp.146-47.

10. Calvert and Kuehn, *Pandora's Box*, p.148.

11. Daniel Bell, *The Coming of Postindustrial Society: A Venture in Social Forecasting* (New York: Basic Books, 1973), p.14.

12. Illich, *Tools for Conviviality*, pp.80-81.

13. John J. Deutsch, "Queen's and the Community," in *Kingston 300: A Social Snapshot* (City of Kingston, 1973), p.98.

14. Kingston Area Economic Development Commission, "Greater Kingston Canada," Kingston, 1991.

15. *The Whig-Standard*, December 1, 1993.

16. See Fred Block, *Revising State Theory: Essays in Politics and Postindustrialism* (Philadelphia: Temple University Press, 1987).

17. Richard Barnet, "The End of Jobs," *Harper's*, September 1993.

18. L. Rosine, "Exposure to Critical Incidents: What Are the Effects on Canadian Correctional Officers?" *Forum on Corrections Research*, 4,1 (March 1992).

19. Author's communication with Robert Fisher, Corrections Canada, Kingston, November 1, 1993.

20. *The Globe and Mail*, January 20, 1994.

21. See, for instance, Sarah Jane Crowe, *Who Cares? The Crisis in Canadian Nursing* (Toronto: McClelland and Stewart, 1991), and Pat Armstrong, Jacqueline Choiniere, and Elaine Day, *Vital Signs: Nursing in Transition* (Toronto: Garamond Press, 1993).

22. Bryan D. Palmer, *Capitalism Comes to the Backcountry: The Goodyear Invasion of Napanee* (Toronto: Between the Lines, 1994), p.66.

23. See Paul Kennedy, *Preparing for the 21st Century* (New York: Random House, 1993), p.89.

24. Ursula Franklin, *The Real World of Technology* (Toronto: Anansi, 1990), ch.1.

25. Lewis Mumford, *The Myth of the Machine*, vol.1., *Technics and Human Development* (New York: Harcourt, Brace and World, 1967); excerpted in D. Miller, ed., *The Lewis Mumford Reader* (New York: Pantheon, 1986), p.318.

26. Michel Foucault, *Discipline and Punish: The Birth of the Prison* (New York: Vintage, 1979), p.169.

27. Franklin, *Real World of Technology*.

28. H. Clare Pentland, *Capital and Labour in Canada, 1650-1860* (Toronto: Lorimer, 1981), p.17.

29. Harris, *Democracy in Kingston*, p.38.

30. James Phelan and Robert Posen, *The Company State: Ralph Nader's Study Group Report on Du Pont in Delaware* (New York: Grossman, 1973), quoted in Philip Matera, *World Class Business: A Guide to the 100 Most Powerful Global Corporations* (New York: Henry Holt, 1992), p.230; for the demise of the UMW in Kingston, see Harris, *Democracy in Kingston*, pp.32-34.

31. B.S. Osborne and D. Swainson, *Kingston: Building on the Past* (Westport, Conn.: Butternut Press, 1988), pp.1, 3.

32. Bryan Palmer, *Working-Class Experience: Rethinking the History of Canadian Labour, 1800-1991*, 2nd ed. (Toronto: McClelland and Stewart, 1992), p.413.

CHAPTER 8

Flexiworkers and the Future:
"All That is Solid Melts into Air"

1. *Maclean's*, April 11, 1994.

2. New Brunswick Departments of Advanced Education and Labour, Income Assistance, "NB Works: A Joint Pilot/Demonstration Project Proposal," Fredericton, 1992, p.1.

3. Ibid., p.2.

4. Government of Canada, "Canadian Paper on New Orientations for Social Policy," presentation to Employment, Labour and Social Affairs Committee (Ministerial Level), Ottawa, December 1992.

5. Special Senate Committee on Poverty, *Poverty in Canada*, pp.ix-x.

6. Margaret Thatcher, *The Downing Street Years* (New York: Harper Collins, 1993), pp.626-67.

7. *Gleaner* (Fredericton), June 16, 1993, quoted in R. Mullaly and J. Weinman, "A Response to the New Brunswick Government's 'Creating New Options,'" mimeo, February 8, 1994.

8. Mullaly and Weinman, "A Response," p.2.

9. Equality Not Charity, "Creating Rebellion Options: A Counter-Discussion Paper," Fredericton, February 1994.

10. *The Economist*, April 6, 1991.

11. Reich, *Work of Nations*, pp.243-51. Reich also pointed out that the U.S. income tax rate for its wealthiest citizens was the lowest in the industrialized world, claiming that while a more progressive system would be no panacea "for widening income disparities rooted in the emerging worldwide division of labour . . . it would at least ameliorate the trend."

12. Human Resources Development Canada, "Improving Social Security in Canada: A Discussion Paper," Ottawa, 1994.

13. Health and Welfare Canada, "Canadian Paper on New Orientations for Social Policy"; CBC Radio News, February 14, 1994.

14. Human Resources Development, "Improving Social Security in Canada," pp.19, 21. A 1980 study of adult basic education in Ontario conducted for the colleges and universities ministry showed a clear statistical relationship between income and participation in adult learning. "Our education system is maintained largely out of public funds. The users of the education system are disproportionately those who are well-off financially. In effect, the poor are subsidizing the rich. This is a fundamental injustice." From "Adult Basic Education and the Educationally Disadvantaged Adult," E. Matthews, 1980, Archives of Ontario, RG 32-30, Interim Access Box 014.

15. R. Morissette, J. Myles, and G. Picot, "What Is Happening to Earnings Inequality in Canada," mimeo, Business and Labour Market Analysis Group, Statistics Canada, 1993.

16. Piore and Sabel, *Second Industrial Divide*, p.17.

17. Ibid., p.262.

18. Harley Shaiken, *Work Transformed: Automation and Labor in the Computer Age* (New York: Holt, Rinehart and Winston, 1984), pp.143, 145.

19. Tom Peters, *Liberation Management: Necessary Disorganization for the Nanosecond Nineties* (New York: Knopf, 1992), pp.312, 314; for an examination of management gurus and what they represent, see Andrzej Huczynski, *Management Gurus: What Makes Them and How to Become One* (New York: Routledge, 1993).

20. Peters, *Liberation Management*, p.xxxi; emphasis in original.

21. Scott Lash and John Urry, *The End of Organized Capitalism* (Oxford: Polity Press, 1987), p.1. See also Karl Marx and Frederick Engels, "Manifesto of the Communist Party," in *Birth of the Communist Manifesto*, ed. Dirk J. Struik (New York: International Publishers, 1971), p.92.

22. For background on Luddism, see E.P. Thompson, *The Making of the English Working Class* (Harmondsworth, England: Penguin, 1963); and Noble, *Progress without People*.

23. David Robertson, "Corporate Training Syndrome," presentation to Conference on Training, Ontario Federation of Labour, October 1990.

24. Dennis K. Williams, "Building a Competitive Work Force," presentation to Training and Education: The Strategic Investment of the 90's, *Financial Post* Conference, November 1990.

25. Quoted in David Robertson and J. Wareham, *Changing Technology and Work: Northern Telecom*, CAW Technology Project, 1989, p.32.

26. Ibid. p.25.

27. Guy Standing, "Alternative Routes to Labour Flexibility," in *Pathways to Industrialization and Regional Development*, ed. M. Storper and A.J. Scott (London: Routledge, 1992). For an analysis of the gendered changes in the Canadian labour force and the union dilemmas, see Ann Duffy and Norene Pupo, *Part-Time Paradox: Connecting Gender, Work and Family* (Toronto: McClelland and Stewart, 1992), especially ch.2, 6.

28. Government of Canada, *The Canadian Pulp and Paper Industry*.

29. Gosta Esping-Andersen, "Labour Movements and the Welfare State: Alternatives in the 1990s" in *Getting on Track: Social Democratic Strategies for Ontario*, ed. Daniel Drache (Montreal and Kingston: McGill-Queen's University Press, 1990).

30. Quoted in Robertson, "Corporate Training Syndrome."

31. Esther Reiter, *Making Fast Food: From the Frying Pan into the Fryer* (Montreal and Kingston: McGill-Queen's University Press, 1991), pp.99, 86, 137.

32. Quoted in Donald Wells, "Are Strong Unions Compatible with the New Model of Human Resource Management?" *Relations Industrielles*, 48,1 (1993).

33. David Crane, *The Next Canadian Century: Building a Competitive Economy* (Toronto: Stoddart, 1992), pp.221-22.

34. Quoted in Peter Marshall, *Demanding the Impossible: A History of Anarchism* (London: Harper Collins, 1992), p.288.

35. Wells, "Are Strong Unions Compatible?"

36. Ibid.

37. Citations from *The Machine That Changed the World* are from Stephen Wood, "The Lean Production Model," presentation to The Lean Workplace Conference, Port Elgin, Ont., September 1993.

38. Peter Warrian, "Case Study: CEWC and Inglis (Cambridge), mimeo, Toronto, 1991.

39. Betcherman, McMullen, Lecki, and Caron, *Canadian Workplace in Transition*, Table 4, p.31.

40. *Inglis: Committed to Excellence*, 109 (November 7, 1991).

41. Larson was quoted in *The Ottawa Citizen*, April 23, 1994. The announcement of the pending closure of the Cambridge plant was in *The Globe and Mail*, November 16, 1994.

42. CAW-Canada, *Work Reorganization: Responding to Lean Production*, Toronto, 1993, p.2.

43. *The Windsor Star*, September 26, 1992.

44. CAW-Canada Research Group on CAMI, *The Cami Report: Lean Production in a Unionized Auto Plant*, Toronto, 1993.

45. *The Globe and Mail*, May 18, 1993.

46. Sam Gindin and David Robertson, "Alternatives to Competitiveness," in *Getting on Track*, ed. Drache, p.33.

47. CAW, *Work Reorganization*, p.12.

48. USWA District 6, *Workplace Reorganization: A Steelworker Guide for Staff and Local Unions*, Toronto, 1992, p.1.

49. CAW, *Work Reorganization*, p.11.

50. Donald Wells, "Lean Production: The Challenge to Labour," presentation to The Lean Workplace Conference, Port Elgin, Ont., September 1993.

51. John O'Grady, "Beyond the Wagner Act, What Then?" in *Getting on Track*, ed. Drache, pp.155-56.

52. Ibid.

53. Ibid., p.158.

54. See Kevin Phillips, *Boiling Point: Democrats, Republicans and the Decline of Middle-Class Prosperity* (New York: Random House, 1993); also his *The Politics of Rich and Poor* (New York: Random House, 1990).

55. Quoted in Gorz, *Critique of Economic Reason*, p.226.

56. P. Kumar, *From Uniformity to Divergence: Industrial Relations in Canada and the United States* (Kingston: IRC Press, 1993), Table 1, pp.12-13.

57. Standing, "Alternative Routes to Labour Flexibility," p.262.

58. Gorz, *Farewell to the Working Class*, p.67.

59. Pentland, *Capital and Labour in Canada*, pp.178-79.

60. Hans Magnus Enzensberger, *Raids and Reconstructions: Essays in Politics, Crime and Culture* (London: Pluto, 1976), p.270.

CHAPTER 9
Work, Time, and the Wheel of Fortune

1. "Solidarity Forever," words reprinted in P.B. Patterson, "Rise Up Singing: The Group-singing Songbook" (Bethlehem, Penn.: A Sing Out Publication, 1988).

2. Judith Timson, "The Four-Day Vacation," *Destinations*, March 1993.

3. Morissette, Myles, and Picot, "What Is Happening to Earnings Inequality in Canada."

4. Michael L. Smith, "Making Time: Representations of Technology at the 1964 World's Fair," in *The Power of Culture: Critical Essays in American History*, ed. Richard Wightman Fox and T.J. Jackson Lears (Chicago: University of Chicago Press, 1993), p.237.

5. Bruce O'Hara, *Working Harder Isn't Working: How We Can Save the Environment, the Economy and Our Sanity by Working Less and Enjoying Life More* (Vancouver: New Star Books, 1993), p.67.

6. Juliet B. Schor, *The Overworked American: The Unexpected Decline of Leisure* (New York: Basic Books, 1991), p.2.

7. Ibid., pp.96-97, 150-51.

8. Mike Cooley, "Work and Time," in *About Time*, ed. Christopher Rawlence (London: Jonathan Cape, 1985), p.33.

9. Ibid.

10. Lewis Mumford, *Technics and Civilization* (New York: Harcourt, Brace, 1934), p.14.

11. Ibid.

12. Jacques Le Goff, *Work and Culture in the Middle Ages* (Chicago: University of Chicago Press, 1980), p.44.

13. Ibid., pp.46-47.

14. Joseph Conrad, *The Secret Agent* (London: J.M. Dent, 1961), pp.33, 35.

15. Michel Foucault, *Discipline and Punish: The Birth of the Prison* (New York: Random House, 1979), p.154.

16. E.P. Thompson, "Time, Work-Discipline and Industrial Capitalism," *Past and Present*, 38 (1967).

17. Quoted in Stephen Kern, *The Culture of Time and Space, 1880-1918* (Cambridge, Mass.: Harvard University Press, 1983), p.110.

18. Statistics Canada, *1992 Household Facilities and Equipment Survey*, Ottawa, 1992.

19. Robert Bellah et al., *The Good Society* (New York: Knopf, 1991), p.93.

20. O'Hara, *Working Harder Isn't Working*, p.16.

21. Statistics Canada, *Initial Data Release from the 1992 General Social Survey on Time Use*, Ottawa, 1993, cited in Vanier Institute for the Family, *Profiling Canada's Families*, Ottawa, 1994.

22. Vanier Institute, *Profiling Canada's Families*, p.107.

23. Raymond Williams, *Keywords: A Vocabulary of Culture and Society* (London: Fontana, 1983), pp.78-79.

24. Schor, *Overworked American*, p.120.

25. Palmer, *Working-Class Experience*, pp.65, 52, 93.

26. Ibid., pp.106-8.

27. Ontario Task Force on Hours of Work and Overtime, *Working Times: The Report of the Task Force*, Toronto, 1987, p.13.

28. Stephen Leacock, *The Unsolved Riddle of Social Justice* (New York: Jonathan Lane, 1920), pp.81-82.

29. Christine Frederick, *Selling Mrs. Consumer* (New York, 1929), p.15, quoted in Ewen, *Captains of Consciousness*, p.171.

30. Quoted in Hunnicutt, "End of Shorter Hours."

31. Leacock, *Unsolved Riddle of Social Justice*, p.28.

32. Herbert Marcuse, *Eros and Civilization* (New York: Vintage, 1962), pp.vii-viii.

33. Benjamin Hunnicutt, "The Pursuit of Happiness," *Context: A Quarterly of Humane Sustainable Culture*, 37 (Winter 1993-94), pp.34-38.

34. See Hunnicutt, "Pursuit of Happiness," and Jamie Swift, "The Brave New World of Work," *CBC-Ideas*, June 29, 1994, which includes an interview with Hunnicutt.

35. Lewis Mumford, *Technics and Civilization* (New York: Harcourt, Brace, 1934), p.22.

36. Charles Taylor, *Sources of the Self: The Making of the Modern Movement* (Cambridge: Cambridge University Press, 1989), p.509.

37. *The Globe and Mail,* October 7, 1994.

38. Gorz, *Farewell to the Working Class,* p.136.

39. Miriam Edelson, "The Boys Just Don't Get It," *Our Times,* October/November, 1994.

40. Williams, *Keywords,* pp.334-37.

41. *The Globe and Mail,* March 13, 1974, quoted in James Rhinehart, *The Tyranny of Work: Alienation and the Labour Process* (Toronto: Harcourt, Brace, Jovanovich Canada, 1987), p.5.

42. Williams, *Keywords,* p.335.

43. Swift, "Brave New World of Work," CBC-*Ideas,* interview with Hunnicutt.

44. Gambling is often referred to by both government officials and its promoters as "gaming." This is perhaps because, despite its popularity and relentless advertising, the idea of gambling has yet to lose all of its residual moral tarnish.

45. David Popenoe, *Private Pleasure, Public Right* (New Brunswick, N.J.: Transaction Books, 1985), cited in Bellah et al., *Good Society,* p.89.

46. Ibid.

47. CBC-Radio News, January 9, 1995.

48. Human Resources Development Canada, *Agenda: Jobs and Growth: Improving Social Security in Canada,* Ottawa, 1994, p.17.

49. *Horizon,* 129-30 (December 1949-January 1950).

50. George Orwell, "James Burnham and the Managerial Revolution," in *The Collected Essays, Journalism and Letters of George Orwell,* vol.4, *In Front of Your Nose* (New York: Harcourt, Brace and World, 1968), pp.160-81.

INDEX

Adams, T. Philip 125
adjustment policy 182
adult education 16; early history of
144-45; incomes and participation
263 n14
Alcan 41-42; employment (1970-89)
50; establishes plant in Kingston
47; wins award, cuts jobs 161
Allmand, Warren, task force report
(1981) 126
Amalgamated Clothing and Textile
Workers Union (ACTW) 195, 203
Americanization 241-42
Antigonish movement 145
apprenticeship 72; call for revival of
79; employer reluctance 91; system
weakening 80, 85-86; lower cost
108; system threatened 88-90
Archer, David 105
automobile industry, 1980 downturn
117; employment 34-35; Windsor
beginnings 24-25; shift in employment patterns 34
Axworthy, Lloyd 122; appointed
minister of human resources
180; green paper on social
security reform 181; review of

social programs 175

banks 4-5, 76-77
Bell, Daniel 159
Benettonism 183
Benson, Herbert 81
Betcherman, Gordon 131
Bevilacqua, Maurizio 17
Birch, David 140
Bondy, Paul 34-35, 38
Border Tool and Die Ltd. 92-93
Bouchard, Benôit 128
Brant Casting 28, 29
Braverman, Harry 80
Brophy, Jim 39, 40
Bureza, Ron 198-99
Burger King 192-93
Burnside, Bob 54-55
Business Council on National Issues
7, 15

CAMI plant 196; strike (1992) 201
Canada Assistance Act 108
Canada Employment and Immigra-
tion Centre (CEIC) 125, 152
Canada Manpower Training
Program (CMTP) 109

Canadian Auto Workers (CAW), Chrysler agreement (1993) 33; education leave and courses 145; Honeywell local 201; membership loss 187; opposition to work reorganization and team concept 201-2; strike at Windsor Casino 38-39; issue of health leave 236

Canadian Forces Base (CFB) Kingston 46; employment (1970-89) 50; as employer 158-59; job loss 231

Canadian General Electric 87-88. See also General Electric

Canadian Manufacturers' Association 65; job training 72, 73; rejects apprenticeship proposal 85; skills shortages 91-92; unemployment 5

Canadian Union of Public Employees 45

Career Development Institutes Ltd. (CDI) 151-54, 155, 156, 157

Cash, Dave 53

casinos. See gambling; gambling industry; Windsor Casino

Chait, Stephen 53

Champion Spark Plug 128; relocates 33

Charters, Meghan 157-59

Chrétien, Jean 124

Chrysler Corporation 34

Clark, Clifford 63

class, attitudes 209-10

clock, in industrialization 217-20

cloth trade 218-19

Communications, Energy, and Paperworkers Union (CEP) 194

Community Industrial Training Committees (CITCs) 148-49, 260 n1

Complax Co. 28-29, 31

computer technology, and flexible economy 183-84; growth in use (1985-91) 15-16, 155; at Kingston job fair 151-52; second industrial revolution 216; training 15-16, 133-35, 156; women users 156. See also computerization

computerization, banks 4-5, 76-77; spurs flexible specialization 183

Concorde 221-22

Connolly, Cyril 243

consumerism 213, 215; as business choice 225; compulsory 222; moulded by market 224-25; as passive culture 229; role of family 222-23; vs. spiritual discipline 230-31; and TV 228

contracting-out, at CFB 158; employee training 152; erodes collective bargaining 189; aids flexibility 106, 185; fragments workforce 191; just-in-time basis 65; management trend 14; as non-standard work 13; as pressure 45; supplements full-time work 182; and unionization 204

Cooley, Mike 216-17

Cooper, Helen 52, 172

corporations, management 64-65; mass production strategy 182-83; mobility 36; skills shortage 123; team strategy 200. See also management

correctional services 12, 162; as employer 159; in Kingston 163-65; Ontario jobs 167; stress 165. See also Corrections Canada; prisons

Corrections Canada, Kingston employment 159; mission statement 164

Crane, David 194

Crispo, John 104

Critical Trade Skills Training 125

Croll, David 26, 108, 177, 236. *See also* Senate Committee on Poverty
Cullen, Bud 124

Dakens, Les 36
Davies, Robertson 48
Deeg, Kurt 37
de Grandpre report 113, 114
delayering 44
Desroches, Dari 19-22
Dewey, John 83
Dixon, Dennis 41-45
Dixon, Heather 41-45, 69
Dodge report (1981) 126
"domestic science" 96. *See also* home economics
Dore, Ronald 139-40
downsizing 189
Drucker, Peter 105
Dunn, Gary 171-72
Du Pont Co. 12; company union 172; employment (1970-89) 50; establishes plant in Kingston 47
Dupuis, Nathan 71-72, 80, 83, 84
Dyer, Lee 63-67
Dylex Corporation 45

Economic Council of Canada 4, 6-7, 257 n17; founded 92, 102-3; *Good Jobs, Bad Jobs* 16; belief in economic growth 110-11
Economic Development Commission. *See* Kingston Area Economic Development Commission; Windsor-Essex County Development Commission
economic growth, and jobs 5-6
economic restructuring, and the clock 207
Edelson, Miriam 236
education, compulsory 82; job training 102-5; role of 70-72; cost of postsecondary 17; home economics 98; "domestic science" 96. *See also* adult education; apprenticeship; informal training; training; vocational education
employers, strengthened position 200-1
employment, conference 147-48; decline of unskilled jobs 123; policy 136; trends in Kingston 50; Shell model 193-94; social policy 206-7; survey (North Kingston) 67. *See also* just-in-time; contracting-out; subemployment
Employment Services Branch (Kingston) 157-58

family income 9
Fisher, Bob 164
flexibility, and attitude adjustment 191-92; to compete 14; deskilling 106, 188; labour 189; as management ideal 122; many meanings 184; in the past 92; of private-sector education 154; in production and organization 64-65; seen as necessary 185; subordinated 190; of workers 106; Xerox scheme 195
flexible manufacturing systems (FMSS) 183-84, 186
flexible specialization 183
food banks 132
Ford, Henry 23, 30; assembly-line production 154; wage increases 32-33
Fordism 183; decline of 190
forest industry 190
Fortin, Judy 129
Foucault, Michel 169-70, 219
Franklin, Benjamin 219
Franklin, Ursula 169, 170, 195
Fromstein, Mitchell 140

Frontier College 144

Galbraith, John Kenneth 11, 67, 176
gambling 24, 239-43, 245, 246;
    known as "gaming" 268 n44
gambling industry, employment 38,
    238-39; Montreal 3; skills training
    116-17. *See also* Windsor Casino
gardening, metaphor vs. gambling
    243-46
General Electric 187. *See also* Cana-
    dian General Electric
General Motors, attitude adjustment
    191; battle with autoworkers (1945)
    25-26; agreement with UAW (1948)
    26; employment 12
Gilbert, Jack 55-58
Goodwin, Jim 166-68
Goodyear (Napanee) 150, 168
Gorz, Andre 6, 207, 235
government, attitude to job training
    81, 83; green paper on social security
    reform 181; industrial strategy goals
    176; purchasing federalism
    approach 107; seat purchases
    (training) 153; shared-cost
    approach 83, 107; spending on
    training 100-1, 108, 109; support
    for private-sector training 157
Gramsci, Antonio 32
Greater Kingston Technology Coun-
    cil 54-55
Gulf & Western 117, 118

Hamilton, Ont. 226
Hawes, Fred 89, 90
Heinz Co. (Leamington, Ont.) 36
"High School Debate" (Kingston)
    70-72
Hobbes, Thomas 6
Hollister, David 39
home economics 98. *See also*

"domestic science"
Hoover, Herbert 227
Houston, Theresa 60, 65
Howe, C.D. 126
Howell, George 72
human resources. *See* labour
Hunnicutt, Benjamin 229
Hurst, Mike (Mayor) 37-38

Illich, Ivan 159
illiteracy 61
immigration 87
Immigration Act (1978) 124
income distribution 39, 159-60;
    polarization 8-9
income tax reform 263 n11
industrial division of labour 73
industrial relations. *See* labour
industrialization 72-73, 75; clock
    217; steam engine 216
industry, attitude to training 81, 108;
    failure to train 90; formal training
    106-7; laissez-faire approach to
    training 112; record in skills train-
    ing 111-12. *See also* management
informal training 88-89; experience
    at Alcan 77-78; government spend-
    ing on 101; of stationary engineers
    105-6; women's 96-97. *See also*
    training
Inglis Co. (Cambridge, Ont.) 197-
    200

Jackson, Brenda and Russ 28-29, 30-
    33
Japan, production model 195-96
job clubs 136-38
job security 5, 65-66, 121; as mirage
    105
job seeking, attitude 150
job training. *See* training
jobless recovery 14

joblessness, link to environmental and cultural issues 215; structured 186. *See also* unemployment
jobsOntario 149-50
just-in-time, employment 65, 122; system for inventory and workers 58

K-D Manufacturing 58-59
kaizen 196
Kellogg Co. 228-29
Kelly, Pat 132
Kelsey-Hayes Co. 25, 33, 132-33
Kent, Tom 102; becomes manpower deputy minister 107; writes throne speech 103; criticizes government approach to labour market 112-13
King, William Lyon Mackenzie 73-74, 82; and immigration 87
Kingston, class divisions 171; early history 45-47; job fair 147-52; hopes for future 12; "new economy" 53-54; sense of place 170, 173; welfare caseload 62-63
Kingston Area Economic Development Commission 12, 50; staffs Queen's bioscience centre 54; R&D 161
Kingston Area Training Advisory Committee 147, 148, 149
Kingston Spinners 231
Klassen-Hayes, Daphne 162-64, 165, 166
Klein, Ralph, introduces "busfare" 178
Knox, Marilyn 69
Kopita, Pat 197-98, 200

labour, control 169-70; empowerment as human resources 193; exports of skilled people 87; imports of skilled people 84, 87; relations with management 198-99; "poaching" 90, 91; polarization 189; postwar accommodation with capital 26-27, 79; postwar accommodation ends 39, 206, 207; decline in working-class jobs 36; need for policy of time 235; relations 65; relations in Windsor 25-26; social policy 234-35; use of term human resources 7, 9. *See also* contracting-out; work
labour market, changes 217; demand balancing supply 92; displacement 60; distribution of jobs 39-40; government spending 149-50; history 6; immigration to meet needs 124; inequality 8-9; needs unserved 234; monetarist approach 112, 124; polarization 10-11, 181-82; supply and training 92; women in 95-97; effect of segmentation on unions 204; Windsor survey 38; winners and losers 242-43
labour movement, activism (1919) 82; attitude to training 81; contradictions 144; crucial challenges 146; defensiveness 143-44; ideals 194; increasing isolation and segmentation 190-91; loss of bargaining power 190; notion of teamwork 194; compared to U.S. and Europe 205-6; level of organization 206; idea of training tax 114; approach to working hours 226-27, 230; shift to service sector 172-73. *See also* unions
labour process, control over 29; deskilling of workers 106. *See also* working conditions; work-time; working hours
Lachapelle, Dave 211-13, 214, 218, 237-38, 246

Lankin, Frances 55
Larson, Peter 200
Lawford, Hugh 56-57, 58
Lawless, Rick 153-54, 156
Le Goff, Jacques 218-19
Leacock, Stephen 226-27
Leake, Albert 84
Leamington, Ont. 36
lean production 195-96; meanings 202-3
learnfare, 178, 180, 234
learning on the job. See informal training
Lecher, Wolfgang 205-6
Lee, Mary Ann 134-35
Lett, Margaret 94-95
Liberal Party (federal), approach to training 102-3
literacy, skills 181; training 176
Lower, Arthur, 47, 48
Luddites 186
lumber workers 207-8

Macdonald Commission 108, 257 n17
Mackasey, Bryce 110
management, control 199; symbolic analysts 65-66; trends 184-85. See also industry; scientific management
Manitoba Technical Training Centre 152
Manpower programs, new approach 109; launched 107
Manser, Garrie 58-62, 168
manufacturing, employment declines 33; deskilling 154-55; growth and decline 3-4; decline in early 1980s recession 50
Marcuse, Herbert 228, 230
Marika 7 221n
market, control of 247
Marten, Peter 179

Marvel training schools 99-100
Mason, Bernie 62-63
Mayo, Elton 80
McKelvey, Bruce 151
McKenna, Frank 174-75; reforms critiqued 178-79; job training 131
McNeill, D.C. 89, 100
McNeill, George 226
McWha, Noella 133-34
Mechanics' Institutes 144
Metro Labour Education Centre (Toronto) 144
Metzler Business School 151
Miller, Donna 149
Miller, Lloyd 201
Miller, Merton 61
Mills, C. Wright 18, 97, 135
Montreal, "new economy" 3
Moore, Marianne 136-37, 140
Mumford, Lewis 169, 217-18, 230
Myles, John 10-11, 13-14, 182

natural resources 3; treatment of 7
neocorporatism 114
New Brunswick, social security reform 175-76
"new economy," in Kingston 53-54; in Montreal 3
New York World's Fair 213-14
Nichols, Paula 51-52
non-standard employment, defined 13; growth of 131; preparation for 176
non-traditional employment 3
Northern Telecom 187-88, 189; attitude adjustment 191-92
nursing 163, 166

O'Grady, John 204
O'Hara, Bruce 215, 223
Oliver, Jake 140-41

on-the-job training. *See* informal training

Ontario government, education program 104; proposal on apprenticeship 85; NDP training policy 114

Organization for Economic Cooperation and Development (OECD), unemployment 204

Orwell, George 246-47; *1984* 1; *Coming up for Air* 1-2, 18, 243

"Orwellian" 1

Payne, Stuart 165

paid educational leave (PEL) 145-46

part-time work, vs. core of full-time workers 182; growth of in services 125; for women 122

Peever, Margaret 116-21, 122, 145

Pentland, Clare 170, 207

Peters, Tom 64, 184-85

Phillips Colleges 157

Pigott, Joseph 85

Piore, Michael 182-84

politicians, attitudes towards 174

Pons, Pam 131, 140, 142-43

postindustrialism 159; notion of 162

poverty, in 1968 108-9; among employed 9, 67

prisons 12; conditions 165; related to culture of compliance 170; in Kingston 47-48. *See also* correctional services; Corrections Canada

private sector, core employees 182; education expansion 157; emphasizes job training 187; training embraced by federal government 124-25. *See also* industry; management

private training business 98-100

professional and technical employees (P&TS) 65-66, 180; workload 210

Progressive Conservative Party, labour policy approach (1980s) 127-28

public sector, growth 108

QL Systems Ltd. 55-58

Queen's University, biosciences centre 54, 162; founded 47; privatizes MBA program 162, 243; R&D 160-61; shapes Kingston 160; School for Policy Studies 63; workforce 50; workplace technology study 15-16, 155-56

Reich, Robert 8, 17, 65, 66, 67, 180

research and development (R&D) 160-61

retraining 154-55. *See also* training

Rhinehart, James 196

Ricardo, David 186

Robarts, John 104

Roosevelt, Franklin 228

Royal Bank of Canada 77

Royal Commission on Industrial Training and Technical Education (1913) 74-76; influence on Technical Education Act 83; recommendations 82; wistful corporatism 93; women's role 95-96

Royal Commission on Tax (1966) 112

Ryan, James 226

Sabel, Charles 182-84

Salomon Brothers 5

Schor, Juliet 215-16, 225

scientific management, deskilling labour 84, 85; division of workforce 80; shifting of production 29

Senate Committee on Poverty (Croll) 108-9; approach to issue 177 service economy, job opportunities 155

service sector, divisions 4; growth of 4, 50-51; increase in jobs 191; women in 97, 98

Shaffer, George 123

Shaiken, Harley 184

Shell Canada (Sarnia) 193-94

Simmons, G.H. 89

Simpkins, Lillian 57

Simpson, Jimmy 81

Skelton, O.D. 63

skills, development 7; shortages 91-92, 123. *See also* training

small business, employment 140; training record 15

Smith, Adam 73

social programs, 1960s initiatives 101-2; offer less protection 18; reform 175-76. *See also* New Brunswick

Sparrow, L.J. 87-88

St. Lawrence College (Kingston) 156; correctional worker course 163-67

Standing, Guy 190, 206-7

Stepping Stones to Employment conference (Kingston) 148-52

Steelworkers. *See* United Steelworkers of America

subemployment 122

Swift, Olive 245-46

Technical and Vocational Training Assistance Act (TVTA, 1960) 100-1; scrapped 107

technical education. *See* vocational education

Technical Education Act (1919) 83, 100

technological change, opposition to 186

technology, control of 169; job displacement 168

telecommunications 6

telemarketing 2

Thatcher, Margaret 177-78

Thompson, E.P. 220-21

time, as social struggle 223-24

timekeeping, implications of 217

Timson, Judith 210

Titanic 221

Toronto-Dominion (TD) Bank 4-5

Total Quality Management (TQM) 196, 201

tourism, as employer 11-12; employment 52; in Kingston 51-52, 159-60, 170; in Windsor 37

Tousignant, Doug 49-51, 52, 77-78

Toyota, approach to training 187; just-in-time inventory system 58

Toyotism 196

Tracy, Brian 151

training, attitude adjustment 191-92; computer courses 15-16, 232; emphasized as solution 113; employee access 156; employer-sponsored 148-49, 152; equity vs. efficiency model 109-10; expectations 245; in Germany and Japan 90; government funding 152; Kingston job fair 147-52; as imperative 14-15, 131; moderation of human behaviour 155; preparation for non-standard jobs 176; polarization 15-16; private-sector failure 113-14; programs in industry (1953) 79; public funding 149-50; aimed at service proletariat 191; tax 114; Thatcher voucher approach 179-80; Toyota approach 187; UIDU funding 127; weakness of private-sector 123. *See also* education; informal training; retraining; skills

Training and Adjustment Board (Ontario) 114

Unemployed Help Centre. *See* Windsor Unemployed Help Centre
unemployment, 1964-69 108; 1980s 5-6; 1981 reports 126; 1981-83 125-26; late 1980s 113; 1989-93 204; consistently high 135; as disciplinary force 235; as enforced idleness 215; as an issue 135; individual responses to 135; job search 62, 139; as personal trouble 139
unemployment insurance (UI), changes 181; cuts 18, 124, 127
Unemployment Insurances Developmental Uses (UIDU) 127
unions, focus on job security 191; influence in Kingston 172; membership decline 204; organizing difficulties 204-5; in public service 48; response to lean production 195; tech-change committees 168; training programs 199; issue of unemployment 143-44; weakness 203; workplace reorganization 193-95, 200-3. *See also* paid educational leave
United Automobile Workers Union, in Windsor 9, 25-26; negotiations with Windsor Bumper 118. *See also* Canadian Auto Workers
United Steelworkers of America 202
UTDC Co. 167, 168

Veley, Mary 231-34
Vickers, Sir Geoffrey 104-5
vocational education 81; courses 86; schools 79; class implications 84; in secondary schools 97-98; commercial and home economics courses 98. *See also* training; vocationalism
Vocational Training Co-ordination Act 85-86
vocationalism 83. *See also* vocational education
Vonnegut, Kurt 102

wages, and standard of living 252 n10
Waldie, Herb 150-51
Walker, Hiram 22-23
Ward, F.W. 99
Watson, John 71-72, 83
Way-Nicholson, Megan 165-66
Weizenbaum, Joseph 58
welfare state, dismantling 17-18
Wells, Donald 194-95, 203
Wickes Windsor Bumper 117-19, 211-12
Wilkinson, Lorraine 9-10
Williams, Dennis 187
Williams, Raymond 224-25
Wilson, Brian James 150
Windsor, early history 22-23; Freedom Festival 115-16; manufacturers' directory 35, 138; ties with Detroit 23-24, 115-16
Windsor Bumper. *See* Wickes Windsor Bumper
Windsor Casino 38-39, 238-39; opening 11; job creation 37. *See also* gambling industry
Windsor-Essex County Development Commission 34, 35
Windsor-Essex Skills Training Advisory Committee, labour-market survey 38
Windsor Unemployed Help Centre (WUHC) 128-34; Business Ventures sessions 140-41; funding 142; federal funding 139; Handling Unemployed Groups (HUG) 142; coping

and life skills 142; programs 144; role in welfare state 143. *See also* job clubs

women, changing role 215-16; and computer technology 156; non-traditional jobs 122; role as housewife 122, 227; at Windsor Bumper 118; and work-time 228-29

work, access to as an issue 211; educational leave 145-46; erosion of jobs 2; issue of caring 234, 236, 244; changes in women's role 215-16; distribution 14; notion of teamwork 193; pressures 214-15; skills unrecognized 232-33; tools 216-17; as vocation 236

work reorganization, individualistic stamp 203. *See also* contracting-out

work-time, control of 218-19; cultural dimension 245; as an issue 216. *See also* subemployment; working hours

worker uprisings 219

workers' control 207-8

Workers' Education Association 144

workfare 178

workforce. *See* labour

working conditions 207; length of work-week 27; overtime 211-12, 218; overwork 215; part-time 122. *See also* working hours

working hours 218-19, 226-27; and family 223-24; Kellogg experiment 228-29; shifting distribution 182. *See also* part-time work

workplace, attitude to 192-93; control 199; democracy 194; deskilling 187-89; quality of 196-97; sexism 120; technology 155-56;

Wyeth Pharmaceuticals 33-34

Xerox Co. 195

Yanovsky, Zal 159-60

Yeates, Maurice 53

youth, aspirations 68-69; and organized labour 146; unemployment 171-72

youth employment 67; symbol of burger-flipper/McJob 192